Categorical Statistics for Communication Research

Categorical Statistics for Communication Research

Bryan E. Denham

WILEY Blackwell

This edition first published 2017
© 2017 John Wiley & Sons, Inc.

Registered Office
John Wiley & Sons, Ltd, The Atrium, Southern Gate, Chichester, West Sussex, PO19 8SQ, UK

Editorial Offices
350 Main Street, Malden, MA 02148-5020, USA
9600 Garsington Road, Oxford, OX4 2DQ, UK
The Atrium, Southern Gate, Chichester, West Sussex, PO19 8SQ, UK

For details of our global editorial offices, for customer services, and for information about how to apply for permission to reuse the copyright material in this book please see our website at www.wiley.com/wiley-blackwell.

The right of Bryan E. Denham to be identified as the author of this work has been asserted in accordance with the UK Copyright, Designs and Patents Act 1988.

Library of Congress Cataloging-in-Publication Data

Names: Denham, Bryan E., 1967– author.
Title: Categorical statistics for communication research / Bryan E. Denham.
Description: Chichester, UK ; Malden, MA : John Wiley & Sons, 2016. | Includes index.
Identifiers: LCCN 2016019969 (print) | LCCN 2016024992 (ebook) |
    ISBN 9781118927106 (cloth : alk. paper) | ISBN 9781118927090 (pbk.) |
    ISBN 9781118927083 (pdf) | ISBN 9781118927076 (epub)
Subjects: LCSH: Communication–Research–Statistical methods. | Statistics.
Classification: LCC P93.7 .D46 2016 (print) | LCC P93.7 (ebook) | DDC 302.23/021–dc23
LC record available at https://lccn.loc.gov/2016019969

A catalogue record for this book is available from the British Library.

Set in 10/12.5pt Galliard by SPi Global, Pondicherry, India
Printed and bound in Malaysia by Vivar Printing Sdn Bhd

1 2017

*For all who aspire to theoretically informed
and methodologically rigorous quantitative research.*

Contents

Detailed Contents

Preface

In June 1946, recognizing an impasse among scientists debating measurement strategies in psychology, S. S. Stevens observed that measurement – the assignment of numerals to objects and events according to rules – depended on the scales to which data were assigned. *Nominal* scales involved the use of numerals as qualitative labels only, "and quite naturally," Stevens (1946, 679) wrote, "there are many who will urge that it is absurd to attribute to this process of assigning numerals the dignity implied by the term measurement." Indeed, simple frequency counts offered limited information, and because advanced analytic techniques for nominal data had not been developed, scholars typically examined relationships two variables at a time, in some cases controlling the level of a third measure.

At the next level of measurement, the *ordinal* level, observations appeared in a ranked sequence. Stevens (1946) cited hardness among minerals as an example, emphasizing that while order did exist, one could not assume equal intervals between observations; the interval between topaz and corundum, for instance, might not equal the interval between corundum and diamond. "In the strictest propriety," Stevens (1946, 679) cautioned, "the ordinary statistics involving means and standard deviations ought not to be used with these scales, for these statistics imply a knowledge of something more than the relative rank-order of the data." In other words, summing a set of scores and dividing by the number of observations could yield a distorted average; the median, or exact middle score, served as a more appropriate measure. Nevertheless, like prominent statisticians who would follow, Stevens did not advocate the wholesale elimination of mean scores at the ordinal level, opting only to state that inaccuracies stood to increase as differences among intervals did the same.

For Stevens, data became "quantitative" at the *interval* level of measurement. Here, means and standard deviations could be computed without qualification, based on assumptions of equal intervals among observations. Centigrade and Fahrenheit temperature scales served as examples of interval measures, to be followed by a fourth and final level of measurement, the *ratio* scale, which contained a point of absolute zero in addition to equal intervals. Periods of time, Stevens wrote, could be measured on a ratio scale, as one could observe a period that was twice as long as another. In contrast, it made little sense to assert that a temperature of 70 degrees Fahrenheit was twice 35-degree weather.

In the years since Stevens (1946) described the four levels of measurement, statisticians have generally referred to data measured at the nominal and ordinal levels as *categorical* while referring to data measured at the interval and ratio levels as *continuous*; the current text focuses on the former. While scholars in social-science fields such as economics, political science, psychology, and sociology have written monographs and longer books addressing the analysis of nominal and ordinal data, communication scholars have lacked a text on which to draw in conducting studies and teaching quantitative research methods. Designed for graduate students in communication as well as faculty members and research professionals in the public and private sectors, *Categorical Statistics for Communication Research* seeks to fill a disciplinary void by presenting communication scholars with a discipline-specific guide to categorical data analysis. In that sense the book seeks to complement statistics texts by Hayes (2005), Reinard (2006), and Weber and Fuller (2013). Their texts contain excellent instruction on techniques such as the analysis of variance (ANOVA) and ordinary least squares (OLS) regression, but the books do not address advanced approaches for analyzing categorical data. In covering advanced techniques in categorical statistics, the present text assumes the reader will have completed an undergraduate course addressing the fundamentals of quantitative research methods. Such a course may have followed one of the texts mentioned above, or perhaps one from Babbie (2015), Keyton (2014), or Wimmer and Dominick (2014).

At the graduate level, communication seminars on quantitative methods tend to focus on techniques that assume interval-level response variables. Following discussions of descriptive statistics and measures of central tendency, instruction often focuses on the t-test and one-way analysis of variance before moving to correlation tests, factorial ANOVA and ordinary least squares regression. Advanced topics include techniques such as structural equations and hierarchical linear modeling (see Hayes, Slater, and Snyder 2008). In contrast, instruction on categorical statistics tends to begin and end with cross-tabulation and chi-square analysis; techniques for the simultaneous analysis of multiple categorical variables often receive little, if any, attention. In addressing such techniques, the current text aspires to the following objectives:

- To provide an accessible guide to the use of categorical statistics, blending necessary background information and formulas for statistical procedures with data analyses illustrating the respective techniques;
- To include examples from multiple areas of the communication discipline;

- To demonstrate how techniques discussed in the book can be applied to data gathered through surveys, content analyses, and other methods;
- To offer useful instructions for categorical data analyses in IBM SPSS®;
- To demonstrate how procedural assumptions – and problems with meeting those assumptions – can offer substantive insight into communication processes;
- To address points of methodological debate in an even-handed manner, identifying approaches within and between areas of study;
- To include a significant number of references for readers seeking additional background information about the techniques addressed.

To meet these objectives, the text begins with an introduction to categorical data analysis, reviewing statistical terminology and the assumptions statisticians have made in developing bivariate and multivariate tests. As the chapter explains, where techniques such as ANOVA and OLS regression assume a normal probability distribution, modeling procedures covered in the current text assume Poisson, binomial, and multinomial distributions, making the techniques comparably robust to non-normal data. Additionally, modeling techniques covered in the text use maximum likelihood estimation (MLE), as opposed to least squares (LSE), in parameterization processes. Because MLE tends to be less biased with large samples (Nunnally and Bernstein 1994), procedures addressed in the book can prove valuable for studies that draw on large public datasets.

Chapter 2 addresses univariate goodness of fit and bivariate tests of independence and association. The chapter focuses on the use of chi-square to assess proportions in the categories of a single variable and independence in contingency tables containing two measures. In doing so, the chapter includes examples from recent content analyses and survey research initiatives, also reviewing measures of association and the likelihood ratio statistic. Regarding terminology, readers may recognize chi-square analysis as a popular nonparametric, "distribution-free" technique for comparing observed and expected frequencies in cross-tabulations (see Conover 1999, Siegel 1956). Absent a point of reference, scholars sometimes regard categorical statistics, in general, as nonparametric; however, as indicated in the previous paragraph, most of the categorical models in this text assume an established distribution. As Anderson and Philips (1981) pointed out, such models focus on parameter estimation and travel beyond mere significance testing. In short, categorical statistics should not be confused with distribution-free, nonparametric procedures such as the Kruskal-Wallis nonparametric ANOVA test or Spearman correlation analysis.

Chapter 3 moves from two-dimensional contingency tables to analyses containing three categorical variables. Analyses of three-dimensional tables involve testing relationships between two measures at a fixed level of a third. As the chapter demonstrates, the Breslow-Day (B-D) and Cochran-Mantel-Haenszel

(C-M-H) tests facilitate comparisons of odds ratios and allow researchers to gather information about three-way tables in an efficient manner. The B-D and C-M-H tests have been applied primarily in studies of health communication, but scholars working in other areas also may find the procedures useful.

Chapter 4 focuses on log-linear modeling, a technique used to examine contingency tables in more than two dimensions. Unlike logit log-linear analysis, addressed in Chapter 5, general log-linear models do not recognize differences between explanatory (independent) and response (dependent) measures; rather, analyses treat all variables as outcomes, modeling the natural logs of cell frequencies. Researchers who use log-linear analysis generally seek to remove parameters from a saturated model, which contains all effects but 0 degrees of freedom, toward a more parsimonious representation of the observed data. Scholars who use logit log-linear models also seek to identify parsimonious relationships, but they do so with a "categorical variable analog" (Knoke and Burke 1980, 25) to ordinary least squares regression. As Chapter 5 explains, logit log-linear models estimate the log odds of a response measure as a function of explanatory variables, and the model also allows more than one dependent variable to be included in a given analysis. In that sense, the logit procedure bears some similarity to the multivariate analysis of variance, which allows more than one response measure to be included in a model.

Chapter 6 addresses binary logistic regression, a technique used in analyses containing a dichotomous dependent variable (e.g., whether or not an individual communicated with an elected representative in the previous 12 months). Logistic regression accommodates categorical and continuous explanatory measures and produces parameter estimates that can be exponentiated to form odds ratios. Chapter 7 covers multinomial logistic regression, which researchers use when a categorical dependent variable contains more than two levels. As an example, scholars of political communication might study predictors of national optimism, with a response measure indicating that survey respondents (a) appeared optimistic about the future of the nation, (b) appeared pessimistic, or (c) appeared neither optimistic nor pessimistic. Although the multinomial procedure treats a response measure as nominal, the technique often proves useful when ordinal logistic regression models, addressed in Chapter 8, do not meet assumptions. As its name implies, the ordinal model analyzes predictors of ordered response measures, which often appear in the form of Likert attitude statements. Researchers may ask study participants to indicate whether they Strongly Agree, Agree, are Undecided, Disagree, or Strongly Disagree that a social protest received fair treatment in the press. While many researchers would treat such a variable as quasi-interval, Likert statements are technically ordinal measures.

Chapter 9 focuses on probit analysis, a technique similar to logistic regression. As the text explains, the binary probit model assumes an underlying, normally distributed, latent continuous measure. This assumption makes the probit model useful in studies involving issues such as gun control, attitudes toward which are more complex than simple for-or-against binaries. Probit analyses contain multinomial and ordinal approaches as well.

Chapter 10 addresses Poisson and negative binomial regression, two techniques used in analyses of count data (i.e., discrete units observed in a given period of time). A communication scholar might use the procedures in studying whether a number of "tweets" posted about a certain topic vary by region of the country and the gender of social media users. If the scholar coded tweets for a subjective measure, such as tone, he or she would need to measure interrater reliability, which the current text covers in Chapter 11. This chapter contains reliability formulas and examples for both nominal and ordinal content variables, explaining how reliability testing advances a study from personal belief to social science, facilitating replication in the process.

In discussing statistical procedures, the text draws on a content analysis published in *Journalism & Mass Communication Quarterly* (Denham 2014) as well as three datasets made available by the Inter-university Consortium for Political and Social Research (ICPSR) at the University of Michigan. The datasets include the 2008 American National Election Study (The American National Election Studies 2008), the 2011 National Survey on Drug Use and Health (United States Department of Health and Human Services 2011), and the 2012 Monitoring the Future study of American youth (Johnston, Bachman, O'Malley, and Schulenberg 2012). Examples illustrate procedures through topics in political and health communication as well as other areas in the communication discipline.

Regardless of the topics communication scholars engage, quantitative research studies invariably contain nominal and ordinal variables. *Categorical Statistics for Communication Research* seeks to enhance the measurement of these variables in statistical systems, contributing both theoretically and methodologically to disciplinary research.

Bryan Denham
July 2016
Clemson, SC

References

Anderson, J. A., and P. R. Philips. 1981. "Regression, Discrimination and Measurement Models for Ordered Categorical Variables." *Applied Statistics*, 30: 22–31.

Babbie, Earl. 2015. *The Practice of Social Research*, 14th ed. Boston, MA: Cengage Learning.

Conover, W. J. 1999. *Practical Nonparametric Statistics*, 3rd ed. New York: John Wiley & Sons, Inc.

Denham, Bryan E. 2014. "Intermedia Attribute Agenda-Setting in the New York Times: The Case of Animal Abuse in U.S. Horse Racing." *Journalism & Mass Communication Quarterly*, 91: 17–37. DOI:10.1177/1077699013514415.

Hayes, Andrew F. 2005. *Statistical Methods for Communication Science*. Mahwah, NJ: Erlbaum.

Hayes, Andrew F., Michael D. Slater, and Leslie B. Snyder, eds. 2008. *The Sage Sourcebook of Advanced Data Analysis Methods for Communication Research*. Thousand Oaks, CA: Sage.

Johnston, Lloyd D., Jerald G. Bachman, Patrick M. O'Malley, and John Schulenberg. 2012. Monitoring the Future: A Continuing Study of American Youth. Funded by National Institute on Drug Abuse. Institute for Social Research, University of Michigan.

Keyton, Joann. 2014. *Communication Research: Asking Questions, Finding Answers*, 4th ed. New York: McGraw-Hill.

Knoke, David, and Peter J. Burke. 1980. *Log-linear Models*. Beverly Hills, CA: Sage.

Nunnally, Jum C., and Ira H. Bernstein. 1994. *Psychometric Theory*, 3rd ed. New York: McGraw-Hill.

Reinard, John C. 2006. *Communication Research Statistics*. Thousand Oaks, CA: Sage.

Siegel, S. 1956. *Non-parametric Statistics for the Behavioural Sciences*. New York: McGraw-Hill.

Stevens, S. S. 1946. "On the Theory of Scales of Measurement." *Science*, 103: 677–680.

The American National Election Studies. 2008. American National Election Study: ANES Pre- and Post-Election Survey. ICPSR25383-v2. Ann Arbor, MI: Inter-university Consortium for Political and Social Research [distributor], 2012-08-30. DOI:10.3886/ICPSR25383.v2.

United States Department of Health and Human Services. 2011. Substance Abuse and Mental Health Services Administration. Center for Behavioral Health Statistics and Quality. National Survey on Drug Use and Health, 2011. ICPSR34481-v2. Ann Arbor, MI: Inter-university Consortium for Political and Social Research [distributor], 2013-06-20.

Weber, Rene, and Ryan Fuller. 2013. *Statistical Methods for Communication Researchers and Professionals*. Dubuque, IA: Kendall Hunt.

Wimmer, Roger D., and Joseph R. Dominick. 2014. *Mass Media Research: An Introduction*, 10th ed. Boston, MA: Cengage Learning.

Acknowledgments

Eight reviewers evaluated the proposal for this text, and I thank each of them for their comments and suggestions regarding structure and content. I also recognize three reviewers who examined the initial draft of *Categorical Statistics for Communication Research*; their feedback helped me to clarify and improve chapter contents, and I very much appreciate their attention to detail.

Elizabeth Swayze provided initial guidance on this project, and I thank her for the support and encouragement. I also thank executive editor Haze Humbert as well as Patrick Wright, who has overseen marketing efforts. I recognize Mary Hall, Julia Kirk, Aneetta Antony, Joanna Pyke, and Roshna Mohan, each of whom contributed to the production of the book. I feel fortunate to have secured a contract with Wiley Blackwell, for in addition to publishing the scholarly journals of the International Communication Association, the company has published leading texts in applied statistics.

I thank the Inter-university Consortium for Political and Social Research (ICPSR) at the University of Michigan for allowing me to demonstrate statistical techniques with data gathered in survey projects such as the American National Election Studies, the National Survey on Drug Use and Health, and the Monitoring the Future study of American youth. I also thank IBM for granting me permission to include screenshots of SPSS for Windows procedures.

Clemson University granted me a sabbatical to pursue this project, and I appreciate that investment in my work. I also acknowledge Clemson Libraries for maintaining electronic subscriptions to key scholarly journals and for retaining classic texts in categorical statistics. I acknowledge my doctoral adviser, M. Mark Miller, PhD, who encouraged me to pursue a minor in applied statistics. In graduate seminars and in dissertation meetings, Dr Miller stressed the importance of theory in quantitative research, and I certainly share his sentiments regarding theoretically informed social science. I also thank Tony Rimmer, PhD, my MA adviser, for introducing me to quantitative research methods at the graduate level.

Finally, I thank students, colleagues, and family members for expressing an interest in the project, offering encouragement, and sending along news items and other materials addressing scientific studies and quantitative research methods.

About the Companion Website

This book is accompanied by a companion website:

www.wiley.com/go/denham/categorical_statistics

The website includes:

- Data files for chapter exercises
- Answers to chapter exercises
- Chapter PowerPoint slides

1

Introduction to Categorical Statistics

This text focuses principally on the analysis of *nominal* and *ordinal* data. Nominal measures contain unordered categories while ordinal variables contain categories in a sequence; both types of measures appear frequently in communication research. At the nominal level, news texts may or may not mention specific issue attributes, and during election years, individuals may or may not view a debate, campaign for a candidate, or vote in a primary. Individuals may be male or female, and they may or may not have served in the military. In addition to these *dichotomous* measures, unordered *polytomous* variables include items such as race, religion, and marital status, each of which contains more than two categories. At the ordinal level, attitude statements frequently include five response options: Strongly Agree, Agree, Undecided, Disagree, and Strongly Disagree. Estimations of risk may range from No Risk to Great Risk, and individuals responding to policy decisions may range from Strongly Approve to Strongly Disapprove in their reactions.

Statistician Alan Agresti (1990) mentioned two additional types of categorical data: *discrete interval* and *grouped interval*. Discrete interval measures often contain a limited number of values, and because they take the form of integers – and integers only – they are not treated as continuous quantitative measures, which can take on any real value. As an example of discrete interval data, a college dean might record the number of people who earn a graduate degree in communication each year, with recipients constituting discrete units. Regarding grouped interval data, researchers sometimes combine continuous interval measures into ordered brackets, as in the case of income, where asking a survey respondent for a specific figure might be considered both invasive and unnecessary. As a second example, while news reports about a given subject might average 731 words, a researcher might be interested in the number of articles that appear in ordered increments of 250 words.

Historical Overview

In covering techniques for analyzing both ordered and unordered categorical variables, the current text recognizes that statisticians have differed in their assumptions and approaches to categorical data analysis. As Powers and Xie (2000) explained, one school of thought considers categorical data part of an underlying continuous distribution, while a second perspective considers categorical data inherently categorical. In historical terms, Agresti (1990) explained that Karl Pearson (1900), who developed the chi-square goodness-of-fit test, assumed continuous distributions underlying categorical variables, while one of Pearson's contemporaries, George Udny Yule (1900), believed that certain types of variables were inherently categorical and did not require assumptions of underlying distributions. Fienberg (2007) observed merit in both perspectives, noting that Pearson and Yule, along with R. A. Fisher (1922a, 1922b), played significant roles in building a foundation for the development of more advanced analytic techniques (see, for additional history, Fienberg and Rinaldo 2007, Plackett 1983). Interestingly, several decades would pass before statisticians developed advanced procedures for categorical data analysis. Most of the modeling techniques covered in the current text emerged after 1960, whereas statisticians had developed multivariate tests for continuous data decades earlier.

Seminal research in communication (e.g., Lazarsfeld, Berelson, and Gaudet 1948) demonstrates how social scientists analyzed and displayed categorical data. Lacking advanced statistical procedures, researchers typically presented data in the form of frequency charts and cross-tabulations. As an example, Table 1.1 contains data gathered in the 1948 election year and published in *Voting: A Study of Opinion Formation in a Presidential Election* (Berelson, Lazarsfeld, and McPhee 1954, 243). The table contains both nominal and ordinal frequency measures and offers descriptive information in a limited but effective manner. Recognizing a pattern between exposure to mass media and level of interest in the presidential election, the authors reported demographic and psychographic information about 814 individuals in Elmira, New York. In the table, numbers appearing in parentheses indicate cell frequencies while figures outside the parentheses indicate the percentage of individuals in each cell who were exposed to media at "High and High-Middle" levels ($N=432$). This approach allowed readers, if so inclined, to calculate the number of respondents in each cell who scored "Low and Low-Middle" on exposure indices ($N=382$), all the while inspecting results across three levels of campaign interest. The use of percentages for "High and High-Middle" media users allowed the authors to

Table 1.1   Example of cross-classifications containing nominal and ordinal measures

	Percentage with High or High-Middle Exposure (on Index)		
		Level of Interest	
Characteristics	*Great Deal*	*Quite a Lot*	*Not Much at All*
(a) Organization Membership:			
Belongs to Two or More	82 (103)	68 (87)	39 (64)
Belongs to One	72 (71)	57 (74)	34 (68)
Belongs to None	62 (100)	47 (112)	24 (126)
(b) Education:			
College	88 (58)	62 (37)	48 (25)
High School	71 (166)	60 (171)	30 (152)
Grammar School or Less	56 (48)	45 (62)	25 (81)
(c) Socioeconomic Status:			
Higher	79 (167)	63 (120)	39 (105)
Lower	60 (108)	52 (153)	25 (154)
(d) Sex:			
Men	72 (122)	60 (124)	38 (110)
Women	71 (153)	54 (149)	25 (149)
(e) Neuroticism:			
Low	77 (112)	64 (106)	30 (100)
High	67 (149)	50 (147)	30 (138)

Note: Table appeared originally in Berelson, Lazarsfeld, and McPhee (1954), *Voting: A Study of Opinion Formation in a Presidential Election*. © 1954 by The University of Chicago. Reprinted with permission, University of Chicago Press.

show statistical patterns that raw cell frequencies would have obscured. Examining the table, one observes that individuals exposed the most to mass media and interested the most in the election belonged to more organizations, had higher levels of education, and appeared in higher socioeconomic classes.

Readers familiar with significance testing may notice that Table 1.1 does not contain chi-square analyses, commonly used to determine whether significant differences exist between observed and expected cell frequencies. Lazarsfeld, a research methodologist, did not consider it appropriate to test bivariate relationships for statistical significance, reasoning that additional variables could alter – or eliminate – significant relationships.[1] As indicated, when Lazarsfeld and his colleagues conducted their election research, multivariate techniques for categorical data had not been developed. For example, log-linear modeling, which examines associations among multiple categorical variables simultaneously, did not exist as such; had the technique been available, Lazarsfeld and other researchers may have used it in analyzing frequency data. In fact, Alwin and Campbell (1987, S147) described log-linear models as, "in many ways, the culmination of the classic Lazarsfeldian tradition. They relate to it directly, rather than obliquely. They focus on tables, the basic building blocks of survey analysis, and they provide precise tests of simple and complex versions of partialling and elaboration hypotheses." Indeed, where Pearson and Yule worked with 2 x 2 contingency tables (i.e., cross-tabulations in which both variables contained two categories), statisticians who developed log-linear models (see Goodman 1978) established approaches for the simultaneous analysis of more than two variables, each of which may have contained more than two categories.

In addition to log-linear modeling, the current text also addresses binary, multinomial, and ordinal logistic regression analyses. As with ordinary least squares (OLS) regression, logistic models examine the effects of one or more independent (explanatory, predictor) variables on a single dependent (response, outcome) measure.[2] Like log-linear models, logistic regression techniques belong to a special class of generalized linear models (GLMs), developed by Nelder and Wedderburn (1972). As explained in Chapter 4 of the current text, a GLM contains a systematic and a random component as well as a link function. Explanatory variables form the systematic component, while a dependent measure and the probability distribution assigned to it constitute the random component (Agresti 2007, 66–67; see also, McCullagh and Nelder 1989). Link functions connect the systematic and random components.

In the case of log-linear and logistic regression techniques, the link function transforms a response measure, such that the dependent variable can be modeled as a linear function of explanatory measures. In OLS regression, a transformation is not necessary, as the procedure models the mean of a dependent variable directly, using an identity link. Log-linear analysis, which models cell frequencies, uses a log link function, while logistic regression analysis, which models a response measure containing a value between 0 and 1 (e.g., a probability), uses the log of the odds. Statisticians who developed logistic regression models (e.g., Cox 1958, McCullagh 1980) built on the work of individuals such as Chester Bliss (1935), who popularized the probit model, and Joseph Berkson (1944), who applied the term *logit* to log odds.[3]

Because advanced modeling techniques for categorical data facilitate the simultaneous examination of multiple variables, they can help to lower the risk of Type I error, or a false rejection of the null hypothesis. When a researcher conducts multiple bivariate analyses using the same set of data, he or she increases the likelihood of identifying "significant" relationships that may be little more than chance occurrences. Yet, legitimate relationships can be rejected when analyses are too conservative; in such cases, Type II error – a failure to reject the null hypothesis when it should be rejected – can occur.[4] As the current text observes, examining multiple variables simultaneously offers an appropriate balance for controlling the two types of error – provided statistical tests meet their assumptions.

Categorical statistics, in general, assume independence among observations, and when that assumption is violated, artificial inflation of a sample may occur, leaving statistical tests technically flawed. The chi-square test statistic, in particular, is sensitive to sample size, and a lack of independence among observations will almost certainly compromise a study. As an example, while a researcher might content analyze 84 individual news reports, a statistician would not consider sentences or paragraphs within those reports independent units. Relatedly, categories within variables should be mutually exclusive and exhaustive, meaning that categories should be independent of one another and contain options for all observations. When categories lack independence and a complete set of response options (or content codes), observations may be classified into more than one category, or no categories at all. In either case, the analysis may not measure what it seeks to measure (i.e., the study may lack internal validity) and attempts to replicate the research may prove futile given an absence of reliability. The following section offers an overview of distributional assumptions and parameter estimation in categorical statistics.

Probability Distributions and Parameter Estimation

A probability distribution links the quantitative outcome of a study with the probability the outcome will occur. In the social sciences, statistics texts focus heavily on outcomes obtained through models such as ordinary least squares regression. OLS regression assumes a *normal* probability distribution with a dependent variable measured at the interval level. It also assumes a random sample and equality of variances, and when analyses meet these assumptions, OLS models yield reliable and parsimonious results. When assumptions are not met, parameters may be misestimated, affecting substantive interpretations (see Aldrich and Nelson 1984).

In contrast to OLS regression, techniques for analyzing categorical response measures vary in the distributions they assume. Models covered in the current text generally assume one of three distributions: *Binomial*, *multinomial*, or *Poisson* (see Plackett 1981). The binomial distribution models the probability of observing a specific number of successes in a certain number of independent trials, and the multinomial distribution models the probability of observing a specific number of successes in each of several categories in a certain number of trials. The Poisson distribution models the probability of observing a specific number of successes in a fixed time period (see also Agresti 2007, 4–16).

In addition to differences in distributional assumptions, categorical procedures rely on a different type of parameter estimation. While OLS regression models contain parameter estimates based on *least squares* (LSE), techniques addressed in this book draw on *maximum likelihood estimation* (MLE). Addressing parameterization, Nunnally and Bernstein (1994, 148) defined an estimator as "a decision rule that results in a particular value or estimate that is a function of the data." Developed by R. A. Fisher (for historical discussion, see Aldrich 1997), MLE selects parameter estimates that have the greatest like-lihood of resulting in the observed sample (Myung 2003). Nunnally and Bernstein noted that while LSE shows little bias in small samples, MLE tends to show greater efficiency and consistency with large datasets.[5]

Example of Maximum Likelihood Estimation

Because MLE is central to procedures addressed in this text, it is important for readers to gain a sense of how maximum likelihood estimates parameters. One approach for demonstrating MLE is to use the binomial formula to first com-pute the probability that, in this case, a certain number of males (y) will appear in a sample (n), with population parameter π indicating the probability of being male. With factorials denoted by !, the binomial formula is expressed as:

$$P(y) = \frac{n!}{y!(n-y)!} \pi^y (1-\pi)^{n-y} \quad y = 0, 1, 2, \ldots, n.$$

To find the probability that three men will appear in a sample of 10 with the probability of male being .50, one would construct the following equation:

$$P(y=3) = \frac{10!}{3!7!} .5^3 (1-.5)^7$$

One would then perform the necessary calculations to arrive at the probability of three men appearing in a sample of 10 individuals, given the .50 probability of being male:

$$P(y=3) = \frac{10 \times 9 \times 8 \times 7 \times 6 \times 5 \times 4 \times 3 \times 2 \times 1}{(3 \times 2 \times 1)(7 \times 6 \times 5 \times 4 \times 3 \times 2 \times 1)} .125(.0078125)$$

$$P(y=3) = \frac{3628800}{(6)(5040)} .125(.0078125)$$

$$P(y=3) = \frac{3628800}{30240} (.000976563)$$

$$P(y=3) = .117$$

In this case, the probability that three men will appear in the sample of 10, given the π value of .50, is 0.117. The formula for the probability distribution and the values of the parameters π and n were known, and the task was to find the probability of observing outcome *y*. But in the practice of quantitative research, parameter values are not known and must be estimated from sample data. A researcher therefore must substitute observed data into the formula for the probability function and then examine different values of π. Using data from the example above, the formula is thus:

$$P\left(y=3\right)=\frac{10!}{3!7!}\pi^{3}\left(1-\pi\right)^{7}$$

After examining the probability for multiple values of π, one arrives at a value for the maximum likelihood estimate; that is, the value of π at which the likelihood of the observed data is highest. Given observed data indicating three successes in 10 independent trials, .3 is the most probable and thus the best estimate for π. Maximum likelihood is used in parameterization processes for advanced categorical statistics and will be referenced throughout the text. The preceding example was designed to familiarize readers with the process, as social scientists often have greater familiarity with least squares estimation (see, for additional discussion, Myung 2003).

A Note on Statistical Software

To facilitate measurement, each chapter in this text contains a section addressing SPSS® techniques for categorical data analysis. Purchased by IBM® in 2009, SPSS is a popular software package in communication and other social science disciplines, and the current text uses SPSS for Windows version 19. Scholars have also used SAS, Stata, and R, each of which functions very well in studies requiring multivariate statistics (see Stokes, Davis, and Koch 2012, Long and Freese 2014). SAS and Stata, in particular, are more powerful than SPSS; however, given the disciplinary prevalence of SPSS, the text focuses on that software. To conserve space in the text, SPSS output is condensed in certain places, with amenable font.

Chapter Summary

This chapter began with examples of categorical variables, noting that statisticians such as Karl Pearson and George Udny Yule differed in their assumptions about measurement. The chapter included an example of cross-classified frequency data from the election research of Berelson, Lazarsfeld, and McPhee (1954) and introduced the types of statistical procedures covered in subsequent chapters. Unlike OLS regression, which assumes a normal probability distribution, procedures covered in this text assume binomial, multinomial, and

Poisson distributions. Additionally, instead of least squares estimation, categorical techniques use maximum likelihood in parameterization processes.

Chapter Exercises

1. Define (or explain) each of the following terms as applicable to categorical statistics.
    a. Binomial distribution
    b. Dependent variable
    c. Dichotomous measure
    d. Discrete interval data
    e. Exhaustiveness
    f. Grouped interval data
    g. Independent variable
    h. Maximum likelihood estimation
    i. Multinomial distribution
    j. Mutually exclusive
    k. Nominal data
    l. Null hypothesis
    m. Ordinal data
    n. Poisson distribution
    o. Polytomous measure
    p. Statistical significance
    q. Type I error
    r. Type II error
2. Classify each measure below as *nominal, ordinal, discrete interval,* or *grouped interval,* briefly justifying each classification.
    a. Position in news organization (advertising representative, editor, publisher, reporter).
    b. Number of "tweets" counted in 60-minute period.
    c. Televised anti-drug spots seen in past week (0, 1–2, 3–5, 6–9, 10–19, 20+).
    d. Attitude toward establishment of federal shield law for journalists (strongly approve, approve, undecided, disapprove, strongly disapprove).
    e. Political ideology (liberal, moderate, conservative).
    f. Political party identification (democrat, republican, independent, other).
    g. Number of violent acts in episode of police drama.
    h. Attention to national television news (no attention, some attention, quite a bit of attention, a great deal of attention).
    i. Empathy for speaker (none, a little, some, a great deal).
    j. Length of public address (less than 60 minutes, 60–74 minutes, 75–89 minutes, 90–104 minutes, 105–119 minutes, 120 or more minutes).

3. Use the binomial formula to find the probability that 4 women will appear in a sample of 10 with the probability of female being .50. Then, calculate a maximum likelihood estimate. Be sure to show your work, indicating the steps taken to perform the calculations.

Notes

1 As readers may recall from previous courses, statistical analyses inform researchers of whether relationships among variables exceed chance occurrences; when relationships exceed chance, they may be considered statistically significant. In communication research, scholars often establish a significance level, or alpha level, of .05, meaning that if researchers conducted the same analysis 100 times, they would observe similar findings in 95 instances. Assuming random sampling, relationships would be attributable to chance just 5 times in 100 cases. Lazarsfeld also recognized that data analysis involves more than testing relationships for statistical significance. Calculating confidence intervals informs researchers of where a true population value is likely to fall given sample estimates (see Koopmans 1987, 226–227). Confidence intervals contain a lower bound and an upper bound, each based on the sample data. In communication research, a confidence coefficient of .95 is typically selected for quantitative studies, and after calculating a confidence interval based on that coefficient, researchers can conclude, with 95% certainty, that a true population value lies between the lower and upper bounds.

2 Because categorical response measures (a) may or may not contain order and (b) may or may not contain more than two categories, scholars who have written texts on categorical data analysis (see, e.g., Agresti 1990, Azen and Walker 2011, Powers and Xie 2000) have focused primarily on dependent variables, whose measurement is critical in drawing valid and reliable conclusions.

3 In general, the current text follows Fienberg (2000, 2007) in discussing advanced techniques for categorical data analysis, observing certain limitations with a GLM approach. As Fienberg (2000, 643–644) explained in the context of log-linear modeling, "It is true that computer programs for GLM often provide convenient and relatively efficient ways of implementing basic estimation and goodness-of-fit assessment. But adopting such a GLM approach leads the researcher to ignore the special features of log-linear models relating to interpretation in terms of cross-product ratios and their generalizations." Fienberg provides additional reasons for approaching statistical procedures as traditionally conceived, noting, for instance, the importance of monitoring empty cells.

4 In recent years communication scholars (see Matsunaga 2007, O'Keefe 2003, 2007, Weber 2007) have debated the advantages and disadvantages of using techniques such as the Bonferroni correction, which divides an established alpha level, or standard for assuming statistical significance (e.g., $p < .05$), by the number of pairwise comparisons. Thus, if a researcher did not use a multivariate technique and instead opted to examine variable relationships using, for instance, three separate cross-tabulations, the Bonferroni correction would require the researcher to divide .05 by three, thereby establishing .0177 as a new significance level.

5   In terms of estimator selection, Nunnally and Bernstein (1994, 154–155) identified four considerations: bias, efficiency, consistency, and sufficiency. Bias indicates the extent to which an expected value or average reflects an actual population parameter, and efficiency indicates the degree to which values obtained from different samples produce similar results. Consistency, Nunnally and Bernstein noted, reveals whether an estimate falls increasingly closer to a population parameter as the size of a sample increases, and sufficiency indicates whether all pertinent sample information is used in parameter estimation.

References

Agresti, Alan. 1990. *Categorical Data Analysis.* New York: John Wiley & Sons.

Agresti, Alan. 2007. *An Introduction to Categorical Data Analysis,* 2nd ed. New York: John Wiley & Sons.

Aldrich, John. 1997. "R. A. Fisher and the Making of Maximum Likelihood." *Statistical Science,* 12(3): 162–176.

Aldrich, John H., and Forrest D. Nelson. 1984. *Linear Probability, Logit, and Probit Models.* Newbury Park, CA: Sage.

Alwin, Duane F., and Richard T. Campbell. 1987. "Continuity and Change in Methods of Survey Data Analysis." *Public Opinion Quarterly,* 51: S139–S155.

Azen, Razia, and Cindy M. Walker. 2011. *Categorical Data Analysis for the Behavioral and Social Sciences.* New York: Routledge.

Berelson, Bernard R., Paul F. Lazarsfeld, and William N. McPhee. 1954. *Voting: A Study of Opinion Formation in a Presidential Campaign.* Chicago: University of Chicago Press.

Berkson, Joseph. 1944. "Application of the Logistic Function to Bio-Assay." *Journal of the American Statistical Association,* 39: 357–365. DOI:10.1080/01621459.1944.10500699.

Bliss, C. I. 1935. "The Calculation of the Dosage-Mortality Curve." *Annals of Applied Biology,* 22: 134–167. DOI:10.1111/j.1744-7348.1935.tb07713.x.

Cox, D. R. 1958. "The Regression Analysis of Binary Sequences." *Journal of the Royal Statistical Society B,* 34: 215–242.

Fienberg, Stephen E. 2000. "Contingency Tables and Log-Linear Models: Basic Results and New Developments." *Journal of the American Statistical Association,* 95: 643–647. DOI:10.1080/01621459.2000.10474242.

Fienberg, Stephen E. 2007. *The Analysis of Cross-Classified Categorical Data,* 2nd ed. New York: Springer.

Fienberg, Stephen E., and Alessandro Rinaldo. 2007. "Three Centuries of Categorical Data Analysis: Log-linear Models and Maximum Likelihood Estimation." *Journal of Statistical Planning and Inference,* 137: 3430–3445. DOI:10.1016/j.jspi.2007.03.022.

Fisher, R. A. 1922a. "On the Interpretation of χ^2 from Contingency Tables, and the Calculation of p." *Journal of the Royal Statistical Society,* 85(1): 87–94.

Fisher, R. A. 1922b. "On the Mathematical Foundations of Theoretical Statistics." *Philosophical Transactions of the Royal Society of London Series A,* 222: 309–368.

Goodman, Leo A. 1978. *Analyzing Qualitative/Categorical Data*. Cambridge, MA: Abt Books.

Koopmans, Lambert H. 1987. *Introduction to Contemporary Statistical Methods*, 2nd ed. Boston: Duxbury.

Lazarsfeld, Paul F., Bernard Berelson, and Hazel Gaudet. 1948. *The People's Choice: How the Voter Makes Up His Mind in a Presidential Election*, 2nd ed. New York: Columbia University Press.

Long, J. Scott, and Jeremy Freese. 2014. *Regression Models for Categorical Dependent Variables Using Stata*, 3rd ed. College Station, TX: Stata Press.

Matsunaga, Masaki. 2007. "Familywise Error in Multiple Comparisons: Disentangling a Knot Through a Critique of O'Keefe's Arguments Against Alpha Adjustment." *Communication Methods and Measures*, 1: 243–265. DOI:10.1080/19312450701641409.

McCullagh, Peter. 1980. "Regression Models for Ordinal Data." *Journal of the Royal Statistical Society B*, 42: 109–142.

McCullagh, P., and J. A. Nelder. 1989. *Generalized Linear Models*, 2nd ed. London: Chapman and Hall.

Myung, In Jae. 2003. "Tutorial on Maximum Likelihood Estimation." *Journal of Mathematical Psychology*, 47: 90–100. DOI:10.1016/S0022-2496(02)00028-7.

Nelder, J. A., and R. W. M. Wedderburn. 1972. "Generalized Linear Models." *Journal of the Royal Statistical Society, Series A*, 135: 370–383.

Nunnally, Jum C., and Ira H. Bernstein. 1994. *Psychometric Theory*, 3rd ed. New York: McGraw-Hill.

O'Keefe, Dan J. 2003. "Colloquy: Should Familywise Alpha be Adjusted? Against Familywise Alpha Adjustment." *Human Communication Research*, 29: 431–447. DOI:10.1111/j.1468-2958.2003.tb00846.x.

O'Keefe, Dan J. 2007. "It Takes a Family – a Well-Defined Family – to Underwrite Familywise Corrections." *Communication Methods and Measures*, 1: 268–273. DOI:10.1080/19312450701641383.

Pearson, K. 1900. "On the Criterion that a Given System of Deviations from the Probable in the Case of a Correlated System of Variables is Such that It Can be Reasonably Supposed to Have Arisen from Random Sampling." *Philosophical Magazine*, Series 5, 50: 157–175.

Plackett, R. L. 1981. *The Analysis of Categorical Data*, 2nd ed. London: Charles Griffin & Co.

Plackett, R. L. 1983. "Karl Pearson and the Chi-squared Test." *International Statistical Review*, 51: 59–72.

Powers, Daniel A., and Yu Xie. 2000. *Statistical Methods for Categorical Data Analysis*. San Diego, CA: Academic Press.

Stokes, Maura E., Charles S. Davis, and Gary G. Koch. 2012. *Categorical Data Analysis Using SAS*, 3rd ed. Cary, NC: SAS.

Weber, Rene. 2007. "To Adjust or Not to Adjust Alpha in Multiple Testing: That is the Question. Guidelines for Alpha Adjustment as Response to O'Keefe's and Matsunaga's Critiques." *Communication Methods and Measures*, 1: 281–289. DOI:10.1080/19312450701641391.

Yule, G. U. 1900. "On the Association of Attributes in Statistics." *Philosophical Transactions of the Royal Society of London*, A194: 257–319.

2

Univariate Goodness of Fit and Contingency Tables in Two Dimensions

More than 100 years after Pearson (1900) introduced the chi-square test, the analytic technique remains one of the most popular approaches for (a) measuring goodness of fit for the categories of a single variable and (b) determining whether an association exists between two variables in a contingency table, or cross-tabulation. This chapter addresses both uses of chi-square and also covers the likelihood ratio statistic as well as exact tests for small samples and McNemar's test for matched pairs. The chapter includes measures of association for both nominal and ordinal variables, addresses frequent pitfalls in bivariate analyses, and concludes with instructions for univariate and bivariate analyses in SPSS.

Chi-Square Test for Goodness of Fit

Research situations sometimes call for comparisons between observed and expected frequencies, the former generated through surveys, content analyses, and other methods and the latter based on theory or existing data. In such studies, the null hypothesis suggests that observed and expected frequencies will be the same (statistically), while the alternative hypothesis anticipates differences, or a lack of fit. Information concerning goodness of fit is useful for both academicians and industry practitioners, and the current chapter applies the chi-square goodness-of-fit test to a situation involving policy compliance at a public university.

In December 2012 *The Red & Black*, the student newspaper at the University of Georgia (UGA), published a feature story on Title IX, a federal law prohibiting sex discrimination in educational institutions that receive federal funding. In the context of competitive sports, Title IX stipulates that percentages of male

Categorical Statistics for Communication Research, First Edition. Bryan E. Denham.
© 2017 John Wiley & Sons, Inc. Published 2017 by John Wiley & Sons, Inc.
Companion Website: www.wiley.com/go/denham/categorical_statistics

Table 2.1   Data for Title IX goodness-of-fit test

Sex	*Number of Athletes*	*Number of Students*	*Percentage*
Male	273	10,218	42.1
Female	266	14,081	57.9

and female athletes on campus reflect overall percentages of male and female students. In her article for *The Red & Black*, Mariana Heredia (2012) included a link to a 2010 university self-study identifying the number of participants in each varsity sport at UGA. Overall, the self-study showed relatively even numbers of male and female athletes; however, the study did not provide enrollment figures. One therefore could not discern whether relatively even numbers of male and female athletes met the proportionality standard of Title IX.

Data available from the United States Department of Education indicate that in 2010, 273 males and 266 females competed in varsity sports at UGA, with 10,218 men and 14,081 women enrolled as undergraduates.[1] One might test the Title IX proportionality standard by applying the chi-square goodness-of-fit test to these athlete and overall enrollment figures, presented in Table 2.1.

The Pearson formula for calculating chi-square (χ^2) is relatively simple: The value of chi-square equals the summation of expected category frequencies subtracted from observed category frequencies, their quantities squared, and then divided by expected frequencies. Thus:

$$\chi^2 = \Sigma \frac{\left(Observed - Expected\right)^2}{Expected}$$

Given the data in Table 2.1, the first task is to calculate expected counts for male and female athletes. In 2010 males accounted for 42.1% of all undergraduate students at the university, and if one multiplies the total number of athletes ($N = 539$) by that percentage, one arrives at an expected count of 227. Subtracting 227 from 539, one arrives at an expected count of 312 female athletes. One can then use the Pearson formula to calculate a χ^2 value for the test statistic:

$$\chi^2 = \frac{\left(273 - 227\right)^2}{227} + \frac{\left(266 - 312\right)^2}{312} = 16.1$$

As indicated in the above calculation, summing the male and female category values results in an overall chi-square value of 16.1. To determine whether this value is statistically significant, one must examine a chi-square table (see Appendix A), which contains values that assume the null hypothesis is true (i.e., that no differences exist between observed and expected frequencies). For the null hypothesis to be rejected, the chi-square value of 16.1 must exceed a corresponding critical value from the table. To locate the critical value, one must calculate degrees of freedom, which Kerlinger and Lee (2000, 231) described as the "latitude of variation contained in a statistical problem." More generally, degrees of freedom refer to the number of items in an equation that are free to vary.

In a chi-square goodness-of-fit test, degrees of freedom are calculated by subtracting 1 from the number of categories. In this case, two categories exist and therefore degrees of freedom equal 1. Looking at the left margin of the chi-square table, the row for 1 degree of freedom shows a critical value of 3.841 for the alpha level of .05; because 16.1 exceeds this value, one rejects the null hypothesis and concludes significant differences between observed and expected frequencies. For the substantive topic of Title IX, one would conclude that even though the numbers of male and female athletes are relatively even, χ^2 shows that their proportions are significantly different.[2]

Chi-Square Test of Independence in Contingency Tables

Contingency tables contain frequency counts for cross-classified categorical variables. Also referred to as cross-tabulations, contingency tables may be used in a descriptive manner, demonstrating statistical patterns through frequency counts and associated percentages, or they may be used to display variables tested for independence with chi-square analysis or the likelihood ratio statistic. In terms of structure, contingency tables contain cells created by intersections among rows and columns. If a table contains two variables, respectively considered explanatory (X) and response (Y), one should position the explanatory measure in the rows and the categories for the response variable in the columns. This allows the reader to look across the table in examining how one measure relates to (or does not relate to) another. If variables are not viewed as explanatory and response, then the choice of row and column measures does not matter as much and is generally based on what a researcher seeks to emphasize or what statistical software packages assume. Although contingency tables often contain multiple rows and columns, the current chapter primarily uses 2×2 tables to demonstrate chi-square and other procedures.[3]

The cross-tabulation in Table 2.2 contains frequency data from a study (Denham 2014) that examined how a 2012 investigative series by *The New York Times* impacted content in other newspapers as well as radio and television news broadcasts. The study, which addressed problems in the sport of horse racing, analyzed the extent to which *The New York Times* transferred the salience of three issue attributes – (a) an injured or deceased horse, (b) equine drug use, and (c) a

Table 2.2 Cross-tabulation of time period by drug-use mentions in horse-racing reports

Time Frame	Equine Drug-Use Mentions in The New York Times		
	Mention	*No Mention*	*Totals*
Before First Investigative Report	*a* 24 (13.6%)	*b* 153 (86.4%)	177
After First Investigative Report	*c* 42 (29.4%)	*d* 101 (70.6%)	143
Totals	66	254	320

trainer suspension or other disciplinary action – to other news organizations. To assess whether an intermedia agenda-setting effect took place, the study analyzed news texts from two periods: prior to publication of the first investigative report and after the first report went to press. As part of examining influence on other news outlets, the study examined patterns of coverage in *The New York Times* itself, reasoning that coverage in other outlets would depend, to some extent, on how the "newspaper of record" portrayed the sport across time.

Table 2.2 indicates that 24 (13.6%) of 177 articles published in *The New York Times* prior to the investigative series mentioned the use of drugs as a story attribute, and following publication of the first investigative report, 42 (29.4%) of 143 news articles mentioned drug use in horse racing. References to equine drug use thus increased from the first to the second period of analysis, but did the references increase beyond chance? The null hypothesis suggested that time frame and mentions of drug use would be independent of one another, while the alternative hypothesis anticipated an association.

As with the goodness-of-fit example, the first step in an independence test is to calculate expected frequencies (i.e., frequencies that assume the null hypothesis is true). This step involves multiplying marginal totals corresponding to each cell and then dividing by the total number of observations. For example, to calculate an expected value for cell *a* in Table 2.2, one would multiply the column total (66) by the row total (177) and then divide by the grand total (320). The process is then repeated for each cell, as indicated below:

$$\text{Cell } a. \ \frac{66 \times 177}{320} = 36.51 \quad \text{Cell } b. \ \frac{254 \times 177}{320} = 140.50$$

$$\text{Cell } c. \ \frac{66 \times 143}{320} = 29.50 \quad \text{Cell } d. \ \frac{254 \times 143}{320} = 113.51$$

After calculating expected cell frequencies, the next step is to calculate a value for chi-square based on the Pearson formula. To begin, one would subtract the expected value for cell *a* (36.51) from the observed value for that cell (24), square the resulting quantity, and divide that number by the expected value. The process is then repeated for each cell in the 2 × 2 cross-tabulation, and summing the cell quantities then provides an overall chi-square value:

$$\text{Cell } a.\ \frac{(24-36.51)^2}{36.51} = 4.29 \quad \text{Cell } b.\ \frac{(153-140.50)^2}{140.50} = 1.11$$

$$\text{Cell } c.\ \frac{(42-29.50)^2}{29.50} = 5.30 \quad \text{Cell } d.\ \frac{(101-113.51)^2}{113.51} = 1.38$$

$$\chi^2 = 4.29 + 1.11 + 5.30 + 1.38 = 12.079$$

For a cross-tabulation, degrees of freedom are calculated by multiplying the number of rows minus one (rows − 1) by the number of columns minus one (columns − 1). In a 2 × 2 contingency table, (2 − 1) multiplied by (2 − 1) equals 1 degree of freedom. Looking to the chi-square table in Appendix A, 12.079 exceeds the chi-square critical value of 3.841 for the alpha level of .05. This means the null hypothesis of independence is rejected and a significant association between time period and references to equine drug use is observed.

Likelihood Ratio Statistic

An alternative to the chi-square test of independence is the likelihood ratio statistic (G^2), which follows a chi-square distribution and usually produces a value close to the chi-square estimate. Because G^2 is the statistic minimized in processes of maximum likelihood estimation (Bishop, Fienberg, and Holland 1975, 125; see also Agresti 1983), it is important to understand how the statistic is calculated. The formula below demonstrates how G^2 is calculated and how it differs from chi-square:

$$G^2 = 2\sum(Observed)\ln\left(\frac{Observed}{Expected}\right)$$

In brief, G^2 is equal to twice the summation of observed values multiplied by the natural log of values derived from dividing observed by expected frequencies. For the data included in Table 2.2, G^2 would be calculated as follows:

$$G^2 = 2\left[24\times\ln\left(\frac{24}{36.51}\right) + 153\times\ln\left(\frac{153}{140.50}\right) + 42\times\ln\left(\frac{42}{29.5}\right) + 101\times\ln\left(\frac{101}{113.51}\right)\right]$$

$$G^2 = 2\left[-10.07 + 13.04 + 14.84 - 11.80\right] = 12.074$$

As indicated, the value for G^2 is often the same as the value for χ^2, and that is basically the case here. Because the likelihood ratio statistic follows a chi-square distribution and is central to modeling techniques addressed in this text, it will be mentioned frequently in subsequent chapters.

Exact Tests for Small Samples

When cells in a cross-tabulation contain relatively few observations (i.e., fewer than five), the chi-square test may be unsuitable as a statistical technique (Upton 1978, 16–18). In such cases, exact tests make it possible to test relationships for significance. Returning to the horse-racing study, Table 2.3 displays a cross-tabulation of time period by mentions of equine drug use, substituting reports from National Public Radio (NPR) for reports published in *The New York Times*. The table contains just 21 observations, and one cell contains a single case. Faced with such a distribution, one might use an exact test, which does not rely on an approximation but rather provides an exact probability level.

Statistician R. A. Fisher (1934) is credited with introducing the exact test for 2 x 2 contingency tables (for discussion, see Agresti 1992). Basing his analysis on marginal totals, Fisher demonstrated how the hypergeometric distribution could explain the probability associated with a specific arrangement of cross-tabulated data. In the formula below, which yields an exact probability level (p), the four cells in Table 2.3 are represented by a, b, c, and d and the grand total is represented by n. Parenthetical expressions in the numerator equal the marginal totals in Table 2.3.

$$p = \frac{(a+b)!(c+d)!(a+c)!(b+d)!}{a!b!c!d!n!} = \frac{9!12!6!15!}{1!8!5!7!21!} = .178$$

Calculation of factorial counts demonstrates that time frame and mentions of equine drug use are independent of one another; that is, there is no association

Table 2.3   Cross-tabulation of time period by drug-use mentions in horse-racing reports on NPR

Time Frame	*Equine Drug-Use Mentions on National Public Radio*		
	Mention	*No Mention*	*Totals*
Before First Investigative Report	a 1 (11.1%)	b 8 (88.9%)	9
After First Investigative Report	c 5 (41.7%)	d 7 (58.3%)	12
Totals	6	15	21

between the two variables. Thus, even though the frequency percentages appear to show differences, none actually exist in the data.

McNemar's Test for Correlated Samples

In some cases, researchers are interested in studying matched pairs. As an example, an experimental researcher may be interested in whether attitudes change before and after exposure to a stimulus. Guided by the binomial distribution, McNemar (1955) developed a version of the chi-square test that analyzes pairs of observations, as opposed to single cases, in a 2 × 2 contingency table.

To illustrate the McNemar procedure, one might consider a hypothetical study addressing fear appeals. A researcher is interested in whether graphic footage of an auto accident will deter adolescents ($N=72$) from texting while driving. The researcher designs an experiment and develops a pretest and posttest instrument for measuring possible changes in attitudes. After conducting the experiment, the researcher creates a cross-tabulation (see Table 2.4) and uses McNemar's test to analyze the data.

In Table 2.4, cell a contains individuals who recognized texting dangers in both the pretest and the posttest, while cell b contains individuals who recognized dangers in the pretest but did not identify dangers in the posttest. Cell c contains individuals who did not recognize dangers in the pretest but did identify them in the posttest, and cell d contains individuals who did not recognize texting dangers in either test. McNemar's test focuses on cells b and c in the cross-tabulation, as cells a and d contain individuals who reported no attitude change.

The null hypothesis anticipates no differences in pretest and posttest attitudes, and as McNemar (1955) showed, that assumption can be tested with the following formula:

$$Q_m = \frac{(b-c)^2}{(b+c)}$$

Table 2.4   Cross-tabulation of paired attitudes

	Recognized Dangers in Posttest	*Did Not Recognize Dangers in Posttest*	*Totals*
Recognized Dangers in Pretest	a 27	b 8	35
Did Not Recognize Dangers in Pretest	c 24	d 13	37
Totals	51	21	72

For the data displayed in Table 2.4, Q_m would be calculated as follows:

$$Q_m = \frac{(8-24)^2}{8+24} = \frac{256}{32} = 8.00$$

In this case, the value for Q_m exceeded the critical value for chi-square with 1 degree of freedom, and the researcher would therefore conclude a significant difference in proportions. It appears the stimulus had an effect on research participants in this hypothetical example, with a significant number of individuals who did not recognize texting dangers in the pretest recognizing dangers in the posttest.

Measures of Association

In his 1967 Presidential Address to the American Statistical Association, Frederick Mosteller (1968, 1) commented on the use of chi-square testing in social science: "I fear that the first act of most social scientists upon seeing a contingency table is to compute chi-square for it. Sometimes this process is enlightening, sometimes wasteful, but sometimes it does not go quite far enough." Applied to data in a contingency table, chi-square indicates *whether* an association exists, and this section of the chapter addresses measures of association, which indicate the *strength* of variable relationships (for reference, see Cliff 1996, Liebetrau 1983). As Mosteller noted, researchers who use chi-square tests sometimes do not provide sufficient information, omitting measures of association and key descriptive statistics.

Odds Ratio

As a measure of association, the odds ratio (OR) is central to statistical procedures such as log-linear modeling and logistic regression analysis. Odds are non-negative and reflect the probability of an event occurring relative to the probability of an event not occurring. An odds ratio measures the association between two odds (Rudas 1998). Data from the cross-tabulation displayed in Table 2.2 can be used to illustrate basic calculations. Prior to the first investigative report published in *The New York Times*, the probability (π) of an equine drug-use mention in news reports was $24/177$, or .136, resulting in the following odds:

$$\frac{\pi}{1-\pi} = \frac{.136}{1-.136} = .157$$

To establish an odds ratio, a researcher would need to calculate the probability of an equine drug-use mention appearing in an article published after the initial investigative report; the probability 42/143, or .294, results in odds of .416. Dividing the odds indicated above (.157) by the odds just calculated (.416), one observes an odds ratio of .377. Thus, the odds of a report from the period before the first investigative article containing an equine drug-use mention were .377 times the odds of a report appearing after the first investigative article making such a reference. If so inclined, one could switch the order of the two odds (i.e., create the ratio .416/.157) and conclude that the odds of an article from the second period containing an equine drug-use mention were 2.2 times the odds of such a mention in an article published prior to the first investigative report.

In an odds ratio, the value 1 is important, as it indicates independence between X and Y. In the example involving equine drug use, dividing odds of .157 by .416 resulted in an odds ratio of .377, which was not close to 1.0. But if one were to replace .157 with .419 and leave the second odds at .416, then the resulting odds ratio would be 1.007, leading the researcher to conclude that drug mentions and time period appeared independent of one another. As Agresti (2007, 29) explained, the value 1 offers "a baseline for comparison."

Conceptualizing categorical data as fixed, Yule (1900, 1912) developed a fundamental cross-product formula for calculating odds ratios:

$$OR = \frac{ad}{bc}$$

Referring to the data in Table 2.2, one would calculate an odds ratio based on the following cross-product equation:

$$OR = \frac{(24)(101)}{(153)(42)} = \frac{2424}{6426} = .377$$

Subsequent use of cross-product ratios in advanced statistical procedures led statisticians such as Fienberg (2007, 5) to consider Yule the "founder" of log-linear modeling, with Knoke and Burke (1980, 10) characterizing odds ratios as the "workhorse" of the log-linear technique.

Odds ratios can prove especially useful in studies containing large samples. In the case of chi-square, nearly any relationship will prove "significant" with enough observations, and therefore a researcher examining relationships in a large-sample contingency table should focus more on association than independence (see Kline 2013). For example, drawing on data gathered in the 2011 National Survey on Drug Use and Health, Table 2.5 contains a 2 × 2 cross-tabulation indicating whether teens who had communicated with parents about drug risks appeared less likely to experiment with marijuana. Chi-square analysis

Table 2.5 Cross-tabulation of communication with parents about drug dangers and experimenting with marijuana

| | *Experimentation with Marijuana* | | |
Communication with Parents	*Experimented*	*Did Not Experiment*	*Totals*
Communicated	*a* 1,824 (16.6%)	*b* 9,189 (83.4%)	11,013
Did Not Communicate	*c* 1,534 (19.3%)	*d* 6,433 (80.7%)	7,967
Totals	3,358	15,622	18,980

Note: These data were gathered in the 2011 National Survey on Drug Use and Health (NSDUH). The data were made available by the Inter-university Consortium for Political and Social Research (ICPSR) at the University of Michigan.

indicates they did χ^2 (1, n = 18,980) = 23.01, $p < .001$ (United States Department of Health and Human Services 2011).

Examining Table 2.5, the difference in marijuana experimentation between adolescents who did and did not communicate with their parents about drugs was less than three percentage points. While 16.6% of those who had communicated had also experimented with marijuana, 19.3% of those who had not communicated had tried the substance. Moving beyond percentage-point differences, the probability of an adolescent who had communicated about drugs experimenting with marijuana was 1,824/11,013, or .166, and the probability of a teen who had not communicated choosing to experiment was 1,534/7,967, or .193. Allowing for the calculation of odds, the odds ratio (.199/.239 = .83) indicates that the odds of individuals who had communicated opting to experiment with marijuana were .83 times the odds of those who had not communicated choosing to experiment. With 1.0 indicating independence, or no association, the variable relationship, identified at $p < .001$ in a chi-square test, did not appear as notable in terms of odds. In fact, as explained in Chapter 3, which includes instructions for calculating 95% confidence intervals for odds ratios, variables are considered independent when confidence intervals include the value 1.0.

Relative Risk

Not to be confused with the odds ratio in a 2 × 2 table is *relative risk*, which compares the probabilities – not the odds – of an event occurring (see Agresti 1990, 17). To calculate relative risk in the data from Table 2.2, a researcher would divide the probability of a drug mention in period one (.136) by the probability of drug mention in period two (.294). The researcher would then report that the probability of a drug mention in period one was .46 times the probability of a drug mention in period two. In general, odds ratios tend to appear more frequently in categorical statistics, largely because techniques such as log-linear modeling and logistic regression analysis produce parameter estimates in the form of log odds, which can be exponentiated to form odds ratios. Scholars such as Zhang and Yu (1998) have proposed approximations of relative risk in the context of logistic regression analysis.

Phi Coefficient

Although this text focuses primarily on odds ratios as measures of association, other approaches do exist for bivariate analyses. In fact, as Van Belle (2002, 7) noted in reviewing 2 × 2 tables, "The number of ways of looking at these simple four numbers is astonishing." Although this chapter does not review all of these approaches, it does discuss those that receive consistent use in social-scientific research, the first being the phi coefficient for 2 × 2 tables (Chedzoy 2006). Yule (1912) proposed the following formula for calculating phi:

$$\Phi = \frac{ad - bc}{\sqrt{(a+b)(c+d)(a+c)(b+d)}}$$

where a, b, c and d apply to four cells in a contingency table. This approach to phi results in values ranging from -1 to $+1$ with 0 indicating no association. Applied to the data in Table 2.2, phi would be calculated in the following manner:

$$\Phi = \frac{2,424 - 6,426}{\sqrt{(177)(143)(66)(254)}} = \frac{-4,002}{\sqrt{424,313,604}} = -.194$$

Examining Table 2.2, the time period following the initial report in *The New York Times* appeared in the second of two row categories, while the lack of a drug mention appeared in the second of two column categories. As indicted by the inverse association, drug mentions appeared more frequently following the investigative report.

The phi coefficient relates to chi-square (see Pearson and Heron 1913) and can be expressed as:

$$\Phi^2 = \frac{\chi^2}{n}$$

where n equals $a + b + c + d$ in a 2×2 contingency table. Given this formula, statistics texts often take a seemingly obvious step for measuring phi directly:

$$\Phi = \sqrt{\frac{\chi^2}{n}}$$

Researchers should take caution when computing phi in this manner, however, as it does not indicate the direction of variable relationships. Again using the data from Table 2.2, phi would be calculated in the following manner:

$$\Phi = \sqrt{\frac{12.079}{320}} = \sqrt{.038} = .194$$

In this calculation, the strength of the relationship is accurate, but the researcher would need to determine its direction. Taking the square root of a chi-square value divided by the number of observations yields a positive value for phi, but does that value agree with how the data are presented in the contingency table? In practical terms, statistical software packages such as SPSS provide a phi value with the positive/negative distinction matching the table a researcher has constructed. Still, it is important to be aware of direction.

Regarding significance testing, because phi relates to chi-square, one can multiply the number of observations in a contingency table by "phi-squared" and arrive at a value for chi-square, which can then be checked against a chi-square distribution. For example, in Table 2.2, if one multiplies an n of 320 by $(.194)^2$, the resulting chi-square value of 12.079 exceeds the critical value of 3.841, indicating a significant relationship between time frame and mentions of drug use in newspaper reports. Again, the researcher would want to double-check the direction of variable relationships.

Cramér's V

In 1946 Harald Cramér extended phi to larger contingency tables. His statistic, Cramér's V, is perhaps the most popular measure of association for nominal variables and can be calculated using the following formula, where k can be either a row or column, whichever is smallest:

$$V = \sqrt{\frac{\chi^2}{n(k-1)}}$$

The value of Cramér's V ranges between 0 and 1, with coefficients closest to 1 indicating comparatively strong relationships. The cross-tabulation in

Table 2.6   Cross-tabulation of race and contacting a public official during previous year

| | *Contacted Public Official to Express Opinion* | | |
Race	*Yes*	*No*	*Totals*
White	247 (18.8%)	1,065 (81.2%)	1,312
Black	60 (11.4%)	465 (88.6%)	525
Other Race	33 (12.9%)	222 (87.1%)	255
Totals	340	1,752	2,092

Note: These data were gathered in the 2008 American National Election Studies (ANES) and were used with the permission of the ICPSR.

Table 2.6, based on data gathered in the 2008 American National Election Studies (ANES), includes three categories of race and an indication of whether or not survey respondents had contacted a public official during the previous year.[4]

To calculate Cramér's V for the data in Table 2.6, a researcher would first need to establish a value for chi-square. In this case, chi-square equals 17.415, and with 2 degrees of freedom, it is significant at $p<.001$. Given the value of chi-square, the number of observations, and the number of columns minus 1, Cramér's V would be calculated as follows:

$$V = \sqrt{\frac{17,415}{(2,092)(2-1)}} = \sqrt{\frac{17,415}{2,092}} = \sqrt{.008423} = .091$$

The value for Cramér's V, .091, indicates that while the cross-tabulation in Table 2.6 showed statistical significance ($p<.001$), the association between race and communicating with a public official during the previous year was actually quite modest.

Pearson's Contingency Coefficient

Pearson also proposed a measure of association for tables larger than 2×2 (see Pearson and Heron 1913), and statistical software packages usually include the Pearson contingency coefficient in cross-tabulation procedures. This measure of association, the maximum value for which depends on the number of rows and the number of columns, can be calculated using the following formula:

$$C = \sqrt{\frac{\chi^2}{n + \chi^2}}$$

Table 2.7   Cross-tabulation of race and perceptions of party most capable of managing economy

| | *Party Most Capable of Managing US Economy* | | | |
Race	*Democrats*	*Republicans*	*Little Difference*	*Totals*
White	495 (35.8%)	338 (24.5%)	549 (39.7%)	1,382
Black	405 (72.2%)	15 (2.7%)	141 (25.1%)	561
Other Race	144 (52.7%)	20 (7.3%)	109 (40.0%)	273
Totals	1,044	373	799	2,216

Note: These data were gathered in the 2008 ANES and were used with the permission of the ICPSR.

The cross-tabulation in Table 2.7, based on data gathered in the 2008 ANES, includes three categories of race along with respondent perceptions of which political party would be most capable of managing the US economy. As with the race measure, the economy variable included three categories, including Democrats, Republicans, and Little Difference (between the two parties). The 3×3 cross- tabulation has $(3-1) \times (3-1) = 4$ degrees of freedom.

To calculate Pearson's C for this contingency table, a researcher would first need to establish a value for chi-square. In this case, given the size of the sample as well as marked differences in category frequency percentages, the chi-square value is quite large: 268.226. This value easily exceeds the critical value in the chi-square distribution, but as indicated by the data in Table 2.6, a large sample requires a measure of association to help clarify the strength of variable relationships. Pearson's C, the significance of which is based on chi-square calculations, serves as an appropriate measure:

$$C = \sqrt{\frac{268.226}{2,216 + 268.226}} = \sqrt{\frac{268.226}{2,484.226}} = \sqrt{.10797} = .329$$

The value of C, .329, indicates a moderate level of association between race and perceptions of the party most capable of managing the US economy, with few individuals apart from White respondents indicating support for the Republican party. Additionally, nearly three in four Black respondents indicated Democrats as the party most capable of managing the economy.

Kendall's Tau

Statistical software packages generally include phi, Cramér's V, and the Pearson contingency coefficient as measures of nominal association, and one of the most popular measures of association for ordinal variables is Kendall's tau_b,

Table 2.8   Cross-tabulation of television and newspaper exposure during 2008 election campaigns

Campaign Programs Watched on TV	Campaign Stories Read in Newspapers			
	A Good Many	*Several*	*Just One or Two*	*Totals*
A Good Many	108 (48.0%)	80 (35.6%)	37 (16.4%)	225
Several	62 (23.8%)	136 (52.3%)	62 (23.9%)	260
Just One or Two	6 (5.8%)	38 (36.5%)	60 (57.7%)	104
Totals	176	254	159	589

Note: These data were gathered in the 2008 ANES and were used with the permission of the ICPSR.

a rank-order measure that can be applied to a pair of ordinal variables or to a two-level nominal measure and an ordinal variable (Agresti 1990).[5] As a measure of association, tau_b ranges between -1 and $+1$ and reflects the proportion of concordant pairs of observations in a contingency table minus the proportion of discordant pairs (Kendall 1945). In statistics, when a subject ranked higher on an X variable also ranks higher on a Y measure, the pair is considered *concordant*. In contrast, a pair is *discordant* when a subject ranked higher on an X measure ranks lower on a Y variable. The following formula, which accounts for ties between X and Y, is used to calculate Kendall's Tau_b:

$$\frac{(C-D)}{\sqrt{\frac{n(n-1)}{2-T_x} \times \frac{n(n-1)}{2-T_y}}}$$

where C is the number of concordant pairs, D is the number of discordant pairs, T_x is the number of pairs tied only on the X variable, and T_y is the number of pairs tied only on the Y variable.

To demonstrate the calculation of Kendall's tau_b, Table 2.8 contains two ordinal variables indicating campaign articles read in newspapers (A Good Many, Several, Just One or Two) and campaign programs watched on television (A Good Many, Several, Just One or Two) by participants in the 2008 ANES study. The table contains fewer observations ($N = 589$) than the previous cross-tabulations, as those who indicated no newspaper exposure and/or TV programs were not included. The 2008 ANES also divided the sample into groups that received "old" and "new" survey questions.

Following Gibbons (1993, 67), the first step in calculating Kendall's tau_b is to establish the total number of concordant and discordant pairs of observations, as illustrated in Table 2.9a.

Table 2.9a   Calculations of concordant and discordant pairs

Pair	Concordant	Totals	Discordant	Totals
1,1	108 (136 + 62 + 38 + 60)	31968	108 (0)	0
1,2	80 (62 + 60)	9760	80 (62 + 6)	5440
1,3	37 (0)	0	37 (62 + 136 + 6 + 38)	8954
2,1	62 (38 + 60)	6076	62 (0)	0
2,2	136 (60)	8160	136 (6)	816
2,3	62 (0)	0	62 (6 + 38)	2728
3,1	6 (0)	0	6 (0)	0
3,2	38 (0)	0	38 (0)	0
3,3	60 (0)	0	60 (0)	0
Totals		55,964		17,938

Table 2.9b   Calculation of corrections for ties

t	$(t^2 - t)/2$	u	$(u^2 - u)/2$
225	25,200	176	15,400
260	33,670	254	32,131
104	5,356	159	12,561
Totals	64,226		60,092

The second step in calculating Kendall's tau_b is to account for ties among observations. Again following Gibbons' (1993, 67) approach, Table 2.9b establishes values for ties among the ranked data, with t representing row totals and u representing column totals.

Given the corrections for ties, one can calculate a value for Kendall's tau_b:

$$\tau_b = \frac{55,964 - 17,938}{\sqrt{\left[589(588)/2 - 64,226\right]\left[589(588)/2 - 60,092\right]}}$$

$$= \frac{38,026}{[330.06][336.26]} = \frac{38,026}{110,985.976} = .34$$

In this case, the value for Kendall's tau_b is .34, indicating a moderate association between exposure to campaign newspaper articles and exposure to television programs about the campaigns. Those who read more newspaper articles also tended to watch more television programs. Conversely, those who read fewer newspaper articles tended to watch fewer TV programs. Such a finding might be important for scholars of political communication, who often seek to determine sources of voter learning during election campaigns.

Goodman and Kruskal's Gamma

A second ordinal measure is Goodman and Kruskal's (1954) gamma coefficient. This measure also calls for calculations of concordant and discordant pairs, with the resulting coefficient, which ranges between −1 and +1, based on the following formula:

$$\gamma = \frac{(C-D)}{(C+D)}$$

Using the concordant and discordant totals established in Table 2.9a, gamma would be calculated in the following manner:

$$\gamma = \frac{(55,964-17,938)}{(55,964+17,938)} = \frac{38,026}{73,902} = .52$$

In this analysis, the gamma coefficient shows a moderately strong association between exposure to newspaper articles and exposure to television programs about the 2008 election campaigns. While statistical software packages display significance levels for the gamma coefficient, special tables are required for traditional (pencil and paper) calculations (Goodman and Kruskal 1980).

Somers' *d*

A third measure of ordinal association is Somers' *d* (Somers 1962). It is included here because of its presence in most statistical software packages and because it allows the researcher to position one variable as independent and one as dependent in calculations. As with gamma, Somers' *d* ranges between −1 and +1 and is interpreted in a similar form. Gibbons (1993, 73) presented the following formula for Somers' *d*, indicating A as the independent variable:

$$d_{B.A} = \frac{2(C-D)}{n^2 - \sum_{i=1}^{r} f_i^2}$$

For the media-exposure cross-tabulation included in Table 2.8, Somers' *d* would be calculated in the following manner, with *n* representing the table total and f_i representing column totals:

$$d_{B.A} = \frac{2(55,964-17,938)}{589^2 - 176^2 - 254^2 - 159^2} = \frac{76,052}{226,148} = .34$$

In this case, with exposure to television programs as the independent variable and exposure to newspaper articles as the dependent measure, Somers' *d* shows an association of .34, indicating a moderate relationship between the two variables.

Points of Concern in Bivariate Analyses

In a classic article addressing mistakes made by researchers who had used chi-square analysis, Lewis and Burke (1949) identified nine sources of error: (1) lack of independence among single events or measures; (2) small theoretical frequencies; (3) neglect of frequencies of non-occurrence; (4) failure to equalize the sum of the observed frequencies and the sum of the theoretical frequencies; (5) indeterminate theoretical frequencies; (6) incorrect or questionable categorizing; (7) use of non-frequency data; (8) incorrect determination of the number of degrees of freedom; and (9) incorrect computations. In 1983, Delucchi followed up on the Lewis and Burke (1949) article and noted that while most of the problems had diminished across time, some of the issues remained. Lack of independence among observations was – and is – a significant point of concern, as Delucchi (1983, 173) noted:

> The value of a chi-square statistic is difficult to evaluate as it is both a function of the truth of the hypothesis under test and a function of sample size. To double the size of a sample, barring sample-to-sample fluctuations, will double the size of the associated chi-square. To compensate for this, the data analyst should always calculate an appropriate measure of association so as to allow for judging the practical, that is, the meaningful significance of the findings.

Because sample size plays an important role in whether chi-square shows significance, researchers should ensure that sample observations are independent of one another; lacking independence, a sample may be subject to artificial inflation and a greater tendency to show "significant" results when none may be present.

Lewis and Burke (1949) also considered problems that arise when cell frequencies are too low. As a test statistic, chi-square is based on a large-sample approximation, and researchers should take caution when applying chi-square tests to tables containing small cell frequencies (see Good and Hardin 2006, 110). Delucchi (1983) recommended that social scientists consider the "Cochran rule." In general, Cochran (1952) suggested, researchers should apply chi-square when no more than 20% of cells have expected frequencies of five or fewer, and no cells contain zero observations.

The sidebar accompanying this section of the chapter addresses the reluctance of research methodologist Paul Lazarsfeld to test bivariate relationships for statistical significance. One of his primary concerns was that introducing additional variables in a statistical system might eliminate significance at the bivariate level, and while the current text does not advocate the elimination of significance testing, it does caution researchers about the myriad of ways in which chi-square, in particular, can indicate significant findings when none may actually exist.

Paul Lazarsfeld, Significance Testing, and Two-Dimensional Contingency Tables

Along with Wilbur Schramm, Harold Lasswell, Kurt Lewin, and Carl Hovland, Paul Lazarsfeld is recognized as a leading figure in the establishment of communication as an academic discipline. Founder of the Bureau of Applied Research at Columbia University, Lazarsfeld specialized in empirical research methods. His panel-study designs set the standard for election research, and his work with Bernard Berelson and Hazel Gaudet (1948) helped to shift the dominant paradigm in media-effects research from the hypodermic-needle model to the two-step flow.

But as much as Lazarsfeld valued quantitative research, he was reluctant to test variable relationships for statistical significance (Rogers 1994). When he and his colleagues conducted their seminal studies of communication processes during election campaigns, technology did not allow for the simultaneous examination of multiple variables. Given the potential influence of "correlated biases" (Selvin 1957) and the realization that bivariate relationships could be altered dramatically by the introduction of a third variable (Lazarsfeld 1961), Lazarsfeld and his colleagues focused more on descriptive statistics.

Social scientists took note of the concerns Lazarsfeld expressed, and in 1970, Denton E. Morrison and Ramon E. Henkel edited *The Significance Test Controversy – A Reader*. In the text, Leslie Kish (1970) addressed the timeless problem of "hunting with a shotgun," explaining that "hunters" typically go in search of significant relationships and, upon locating them, attempt to ascribe meaning to what are, in many cases, chance occurrences. With enough two-dimensional analyses and a sufficiently large sample, many relationships will appear "significant."

In his 1990 text *Survey Research Methods*, Earl Babbie suggested that significance testing, while based on sound logic, could not be justified in certain conditions. "Tests of significance make sampling assumptions that are virtually never satisfied by actual sampling designs," Babbie (1990, 302) wrote. Significance tests also assume nonsampling errors and are frequently applied to associations that violate basic assumptions (e.g., Pearson correlation coefficients based on ordinal measures). Lastly, Babbie argued, researchers frequently misinterpret statistical significance as an indicator of the strength of an association.

More recently, in communication, Timothy R. Levine and his colleagues (2008a, 2008b) discussed issues associated with significance testing in quantitative research. Their papers addressed issues such as sample size, power, and associated error. To minimize problems associated with significance testing, Levine and his colleagues recommend reporting effect sizes and confidence intervals, among other approaches (see also Levine 2013). Overall, while communication scholars should not abandon testing for significance, they should inform readers of additional information in reporting results (Kline 2013).

SPSS Analyses

This chapter has addressed goodness of fit, independence, and association, and it now offers an overview of SPSS procedures, beginning with the chi-square goodness-of-fit test. In demonstrating this test, the chapter draws on the Title IX data from Table 2.1. The chapter then uses the horse-racing data from Table 2.2 to demonstrate the chi-square test of independence.

Testing Goodness of Fit in SPSS

Readers may recall from the Title IX data in Table 2.1 that expected counts of male and female athletes differed significantly from observed numbers χ^2 $(1, n = 539) = 16.1, p < .001$. This section of the chapter uses the Title IX data to demonstrate goodness-of-fit testing in SPSS. Because the present test is based on total frequency counts (as opposed to raw data), preliminary weighting is required. Observing Figure 2.1, a researcher can assign variable weights through the following steps:

- Open the Data menu.
- Select Weight Cases.
- Select the Frequency Variable to be weighted (Count) and click OK.

SPSS will then recognize 273 males and 266 females in the system.

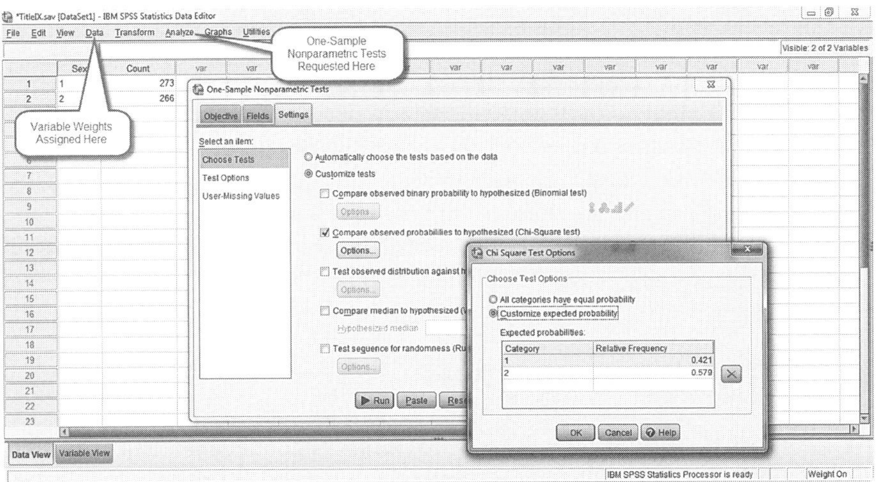

Figure 2.1 Display of SPSS Goodness-of-Fit windows. Source: SPSS® Reprints Courtesy of International Business Machines Corporation, © 2014 International Business Machines Corporation

Hypothesis Test Summary

	Null Hypothesis	Test	Sig.	Decision
1	The categories of Sex occur with the specified probabilities.	One-Sample Chi-Square Test	.000	Reject the null hypothesis.

Asymptotic significances are displayed. The significance level is .05.

Figure 2.2 Display of SPSS Goodness-of-Fit results. Source: SPSS® Reprints Courtesy of International Business Machines Corporation, © 2014 International Business Machines Corporation

With preliminary weighting complete, goodness of fit can be tested. Again observing Figure 2.1, a researcher can compare observed and expected frequencies through the following steps:

- Open the Analysis menu.
- Select Nonparametric Tests followed by One Sample.
- In the Objective section, select Customize analysis.
- In the Fields section, select a measure (Sex) for the Test Field.
- In the Settings section, select Compare observed probabilities to hypothesized, and in Options, choose Customize expected probability. (Recalling the data in Table 2.1, the expected probability for males is .421 and the expected probability for females is .579).
- Select OK in Chi-Square Test Options.
- Select Run in One-Sample Nonparametric Tests.

As illustrated in Figure 2.2, the goodness-of-fit test produced output re-stating the null hypothesis and indicating a decision in reference to it. In this case, SPSS recommended rejection of the null hypothesis, identifying significant differences between observed and expected frequencies. (Note: The SPSS syntax for conducting a goodness-of-fit test is presented in Appendix B.)

Having demonstrated the goodness-of-fit procedure in SPSS, the chapter now turns to the chi-square test of independence.

Testing Independence in SPSS

As indicated in Figure 2.3, testing two variables for independence in SPSS requires the researcher to first create a cross-tabulation. To do so:

- Open the Analyze menu.
- Select Descriptive Statistics and then Crosstabs.

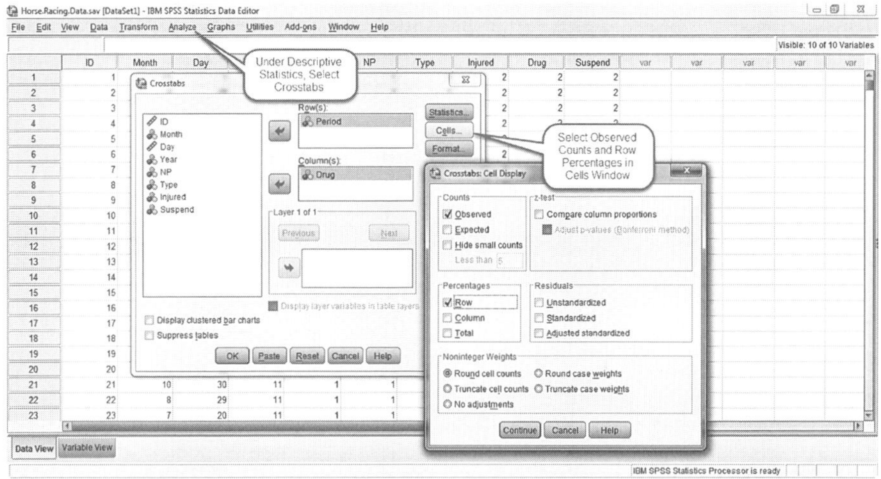

Figure 2.3 Display of SPSS Crosstabs and Cells windows. Source: SPSS® Reprints Courtesy of International Business Machines Corporation, © 2014 International Business Machines Corporation

- In the Crosstabs window, enter a Row (explanatory) and a Column (response) variable.
- Click Cells, select Observed Counts and Row Percentages, and click Continue. (While SPSS can also generate Expected Counts as well as Column Percentages, those options need not be selected in every case, as they can lead to confusion and errors in reporting results.)

As indicated in Figure 2.4, when a cross-tabulation has been created and cell information specified, a researcher can then select appropriate statistics. To do so:

- Select Statistics.
- Select Chi-square and Phi and Cramer's V (for nominal data).
- Click Continue in Statistics.
- Click OK in Crosstabs.

Before reviewing statistical output, other measures of association available in SPSS should be recognized. As shown in Figure 2.4, nominal measures include the Contingency coefficient, Lambda, and the Uncertainty coefficient.[6] Also of note in Figure 2.4 are four ordinal measures of association: Gamma, Somers' d, Kendall's tau$_b$, and Kendall tau$_c$. Additional options will be covered in Chapter 3.

Table 2.10 contains SPSS output for the cross-tabulation of Period across Drug (use) for 320 reports published in *The New York Times*. Consistent with Table 2.2, the cross-tabulation reveals that 13.6% of articles published prior to

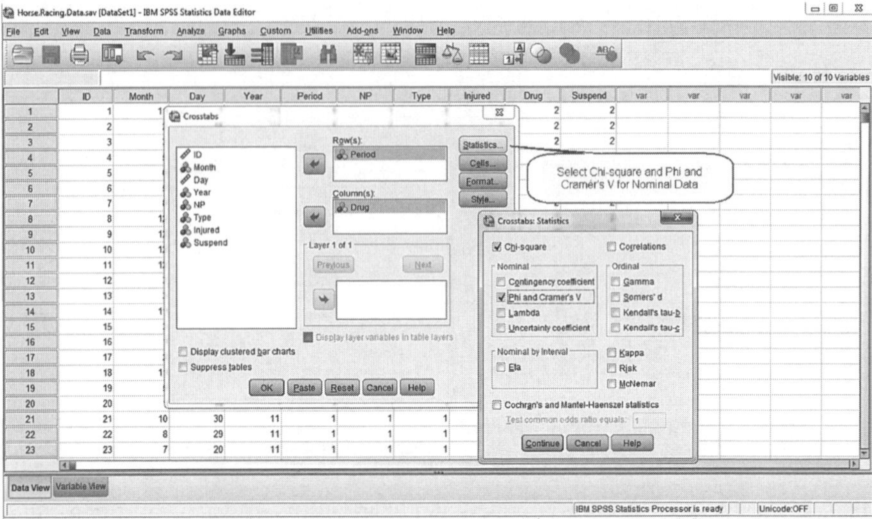

Figure 2.4 Display of statistics available in SPSS Crosstabs procedure. Source: SPSS® Reprints Courtesy of International Business Machines Corporation, © 2014 International Business Machines Corporation

the first investigative report mentioned drug use, while 29.4% of articles published after the first article mentioned equine drugs. As indicated in the table, the Pearson chi-square test proved significant, as it did earlier, along with the significant likelihood ratio statistic. The table also contains output for Fisher's exact test, which is not necessary in this case, as well as a continuity correction and a linear-by-linear association.[7] The table shows measures of association for the 2×2 table, with the phi coefficient at $-.194$, consistent with earlier calculations. While the output also includes a value for Cramér's V, it is not necessary in this case.

A Note on Style

Prominent journals in communication tend to follow guidelines set forth in the *Publication Manual of the American Psychological Association* (6th edition, 2013). In text, a researcher would report the statistics shown in Table 2.10 in the following manner: $\chi^2 (1, 320) = 12.079$, $p < .001$, $\Phi = -.194$. Consistent with the advice of Kerlinger and Lee (2000), a researcher would also want to mention key descriptive statistics to assist the reader in following the quantitative results.

Table 2.10 Display of SPSS cross-tabulation and chi-square statistics

Period * Drug Cross-tabulation

			Drug		
			Mention	No Mention	Total
Period	Before	Count	24	153	177
		% within Period	13.6	86.4	100.0%
	After	Count	42	101	143
		% within Period	29.4%	70.6%	100.0%
	Total	Count	66	254	320
		% within Period	20.6%	79.4%	100.0%

Chi-Square Tests

	Value	df	Asymp. Sig. (2-sided)	Exact Sig. (2-sided)	Exact Sig. (1-sided)
Pearson Chi-Square	12.079	1	.001		
Continuity Correction	11.132	1	.001		
Likelihood Ratio	12.074	1	.001		
Fisher's Exact test				.001	.000
Linear-by-Linear Assoc.	12.041	1	.001		
N of Valid Cases	320				

Symmetric Measures

		Value	Approx. Sig.
Nominal by Nominal	Phi	−.194	.001
	Cramér's V	.194	.001
N of Valid Cases		320	

Chapter Summary

This chapter began by addressing chi-square analysis as a test for (a) goodness of fit and (b) independence. It covered the likelihood ratio statistic as well as exact tests for small samples and McNemar's test for matched pairs before moving to measures of association. The chapter addressed odds ratios and the calculation of nominal measures such as phi, Cramér's V, and Pearson's contingency coefficient before addressing ordinal measures Kendall's tau_b, Goodman and Kruskal's gamma, and Somers' *d*. A section addressed points of concern in bivariate analyses and cautioned researchers about significance testing. Lastly, the chapter offered an overview of SPSS techniques for analyzing goodness of fit, independence, and association.

Chapter Exercises

1. Research in sports communication has found that, during broadcasts of elite sporting events, commentators sometimes focus disproportionately on the physical appearance of female athletes, diminishing the abilities and achievements of the competitors (see Billings, Halone, and Denham 2002). In light of such patterns, a broadcast executive has commissioned a (hypothetical) study in which a researcher will compare descriptors of female athletes in a recently televised tournament with descriptors from coverage the previous year. The data appear below.

Descriptors	*Previous Percentage of Descriptors*	*Current Observed Frequencies*
Speed	15	23
Strength	16	34
Agility	17	36
Composure	14	30
Appearance	14	18
Intelligence	12	24
Other	12	20
Totals	100%	185

    Given the previous percentages, calculate expected frequencies for the current data and use the chi-square goodness-of-fit test to determine whether significant differences between observed and expected frequencies appeared. Remember that degrees of freedom will need to be adjusted based on the number of categories in the equation. What do the results suggest about descriptor patterns relative to the findings of earlier research?

2. Tables 2.2 and 2.10 displayed data gathered from horse-racing reports published in *The New York Times*. Referring to Table 2.2, substitute the numbers 6, 48, 14, and 38 in cells *a*, *b*, *c*, and *d*, respectively. These numbers were derived from content in the *Washington Post*, as opposed to *The New York Times*, and also pertained to time period and drug mentions. After setting up a cross-tabulation, calculate the following: (a) chi-square statistic; (b) likelihood ratio statistic; (c) odds ratio; and (d) phi coefficient. Provide both statistical and substantive interpretations of your results.

3. Table 2.8 contained two ordinal measures, one addressing campaign programs watched on television during the 2008 election year and the other focusing on campaign stories read in newspapers. The cross-tabulation below contains a variable addressing campaign articles read in magazines in place of the television variable. Given this cross-tabulation, calculate the following: (a) chi-square statistic; (b) likelihood ratio statistic; (c) Kendall's

tau$_b$, (d) Goodman and Kruskal's gamma, and (e) Somers' *d*. Provide both statistical and substantive interpretations of your results.

Campaign Stories Read in Magazines	Campaign Stories Read in Newspapers			
	A Good Many	*Several*	*Just One or Two*	*Totals*
A Good Many	52 (76.5%)	12 (17.6%)	4 (5.9%)	68
Several	36 (24.2%)	93 (62.4%)	20 (13.4%)	149
Just One or Two	22 (19.1%)	42 (36.5%)	51 (44.3%)	115
Totals	110	147	75	332

Note: These data were gathered in the 2008 ANES and were used with the permission of the ICPSR.

Notes

1 The website of the United States Department of Education (http://ope.ed.gov/ athletics/) contains the Equity in Athletics Data Analysis Cutting Tool, which provides customized reports of male and female athletic participation at colleges and universities in the United States.
2 This example is provided solely to illustrate a statistical technique. Many universities meet Title IX expectations by demonstrating they are moving in the direction of proportionality.
3 It should be noted here that chi-square is a continuous distribution, and counts in contingency tables come from discrete distributions such as the binomial. The χ^2 value is therefore an approximation. While statisticians frequently use 2×2 tables to demonstrate calculations, chi-square sometimes offers a poor approximation in a 2×2 table, which contains 1 degree of freedom. Historically, statisticians have used Yates' (1934) continuity correction for 2×2 tables (see also Grizzle 1967), and modern software packages offer continuity corrections as well as exact tests to confirm findings.
4 As a matter of protocol, it should be noted that in 2008 the ANES oversampled African Americans and Latinos. The survey also included new versions of questions already in use, in some cases splitting the sample into two sets of respondents. For additional information about sampling techniques in 2008, see http://www.election studies.org/studypages/2008prepost/2008prepost.htm.
5 Kruskal and Wallis (1952) developed a nonparametric one-way analysis of variance procedure to test for equality of distributions, and their test may prove useful when one measure is nominal and the other ordinal.
6 Lambda and the uncertainty coefficient are measures of association that indicate a proportional reduction in error when one variable is used to predict another. Additional information on these measures can be found in the manual *IBM SPSS Statistics Base 20* (2011).

7   The continuity correction should be examined in 2 × 2 tables, especially in cases when one cell has fewer than 5 observations. A linear-by-linear association measures linearity between ordinal measures. As with Note 6, additional information on these measures can be found in the manual *IBM SPSS Statistics Base 20* (2011).

References

Agresti, Alan. 1983. "A Survey of Strategies for Modeling Cross-Classifications Having Ordinal Variables." *Journal of the American Statistical Association*, 78: 184–198.

Agresti, Alan. 1990. *Categorical Data Analysis*. New York: John Wiley & Sons.

Agresti, Alan. 1992. "A Survey of Exact Inference for Contingency Tables." *Statistical Science*, 7: 131–153.

Agresti, Alan. 2007. *An Introduction to Categorical Data Analysis*, 2nd ed. Hoboken, NJ: John Wiley & Sons.

American Psychological Association. 2013. *Publication Manual of the American Psychological Association*, 6th ed. Washington, DC: American Psychological Association.

Babbie, Earl. 1990. *Survey Research Methods*, 2nd ed. Belmont, CA: Wadsworth.

Billings, Andrew C., Kelby K. Halone, and Bryan E. Denham. 2002. "'Man, That was a Pretty Shot': An Analysis of Gendered Broadcast Commentary of the 2000 Men's and Women's NCAA Final Four Basketball Championships." *Mass Communication & Society*, 5: 295–315. DOI:10.1207/S15327825MCS0503_4.

Bishop, Yvonne M. M., Stephen E. Fienberg, and Paul W. Holland. 1975. *Discrete Multivariate Analysis: Theory and Practice*. Cambridge, MA: MIT Press.

Chedzoy, O. B. 2006. "Phi-Coefficient." *Encyclopedia of Statistical Sciences*. New York: John Wiley & Sons.

Cliff, Norman. 1996. "Answering Ordinal Questions with Ordinal Data Using Ordinal Statistics." *Multivariate Behavioral Research*, 31: 331–350. DOI:10.1207/s15327906mbr3103_4.

Cochran, William G. 1952. "The χ^2 Test of Goodness of Fit." *The Annals of Mathematical Statistics*, 23: 315–345.

Cramér, Harald. 1946. *Mathematical Methods of Statistics*. Princeton, NJ: Princeton University Press.

Delucchi, Kevin L. 1983. "The Use and Misuse of Chi-Square: Lewis and Burke Revisited." *Psychological Bulletin*, 94: 166–175.

Denham, Bryan E. 2014. "Intermedia Attribute Agenda-Setting in the New York Times: The Case of Animal Abuse in U.S. Horse Racing." *Journalism & Mass Communication Quarterly*, 91: 17–37. DOI:10.1177/1077699013514415.

Fienberg, Stephen E. 2007. *The Analysis of Cross-Classified Categorical Data*, 2nd ed. New York: Springer.

Fisher, R. A. 1934. *Statistical Methods for Research Workers*, 5th ed. Edinburgh: Oliver and Boyd.

Gibbons, Jean Dickinson. 1993. *Nonparametric Measures of Association*. Newbury Park, CA: Sage.

Good, Phillip I., and James W. Hardin. 2006. *Common Errors in Statistics (and How to Avoid Them)*. Hoboken, NJ: John Wiley & Sons.

Goodman, L. A., and E. H. Kruskal. 1954. "Measures of Association for Cross-Classifications." *Journal of the American Statistical Association*, 49: 732–764.

Goodman, L. A., and W. H. Kruskal. 1980. *Measures of Association for Cross-Classifications*. New York: Springer.

Grizzle, James E. 1967. "Continuity Correction in the χ^2 Test for 2 × 2 Tables." *The American Statistician*, 21: 28–32.

Heredia, Mariana. 2012. "Title IX Compliance: A Tricky Game for the University." *The Red & Black*, December 3, 2012.

International Business Machines Corporation (IBM). 2011. *IBM SPSS Statistics Base 20*. Raleigh, NC: IBM.

Kendall, M. G. 1945. "The Treatment of Ties in Rank Problems." *Biometrika*, 33: 239–251.

Kerlinger, Fred N., and Howard B. Lee. 2000. *Foundations of Behavioral Research*, 4th ed. Belmont, CA: Wadsworth.

Kish, Leslie. 1970. "Some Statistical Problems in Research Design." In Denton E. Morrison and Ramon E. Henkel (Eds), *The Significance Test Controversy – A Reader* (pp. 127–141). Chicago: Aldine Publishing Company.

Kline, Rex B. 2013. *Beyond Significance Testing: Statistics Reform in the Behavioral Sciences*, 2nd ed. Washington, DC: American Psychological Association.

Knoke, David, and Peter J. Burke. 1980. *Log-linear Models*. Beverly Hills, CA: Sage.

Kruskal, W. H., and W. A. Wallis. 1952. "Use of Rank in One-Criterion Variance Analysis." *Journal of the American Statistical Association*, 47: 401–412.

Lazarsfeld, Paul F. 1961. "The Algebra of Dichotomous Systems." In Herbert Solomon (Ed.), *Studies in Item Analysis and Prediction* (pp. 111–157). Stanford, CA: Stanford University Press.

Lazarsfeld, Paul F., Bernard Berelson, and Hazel Gaudet. 1948. *The People's Choice: How the Voter Makes Up His Mind in a Presidential Election*, 2nd ed. New York: Columbia University Press.

Levine, Timothy R. 2013. "A Defense of Publishing Nonsignificant (ns) Results." *Communication Research Reports*, 30: 270–274. DOI:10.1080/08824096.2013.806261.

Levine, Timothy R., René Weber, Craig Hullett, Hee Sun Park, and Lisa L. Massi Lindsay. 2008a. "A Critical Assessment of Null Hypothesis Significance Testing in Quantitative Communication Research." *Human Communication Research*, 34: 171–187. DOI:10.1111/j.1468-2958.2008.00317.x.

Levine, Timothy R., René Weber, Hee Sun Park, and Craig R Hullett. 2008b. "A Communication Researchers' Guide to Null Hypothesis Significance Testing and Alternatives." *Human Communication Research*, 34: 188–209. DOI:10.1111/j.1468-2958.2008.00318.x.

Lewis, Don, and C. J. Burke. 1949. "The Use and Misuse of the Chi-Square Test." *Psychological Bulletin*, 46: 433–489.

Liebetrau, Albert M. 1983. *Measures of Association*. Newbury Park, CA: Sage.

McNemar, Q. 1955. *Psychological Statistics*. New York: John Wiley & Sons.

Morrison, Denton E., and Ramon E. Henkel. 1970. *The Significance Test Controversy – A Reader*. Chicago: Aldine Publishing Company.

Mosteller, Frederick. 1968. "Association and Estimation in Contingency Tables." *Journal of the American Statistical Association*, 63: 1–28.

Pearson, Karl. 1900. "On the Criterion that a Given System of Deviations from the Probable in the Case of a Correlated System of Variables is Such that It Can be Reasonably Supposed to Have Arisen from Random Sampling." *Philosophical Magazine*, Series 5, 50: 157–175.

Pearson, Karl, and David Heron. 1913. "On Theories of Association." *Biometrika*, 9: 159–315.

Rogers, Everett M. 1994. *A History of Communication Study: A Biographical Approach*. New York: The Free Press.

Rudas, Tamas. 1998. *Odds Ratios in the Analysis of Contingency Tables*. Thousand Oaks, CA: Sage.

Selvin, Hanan C. 1957. "A Critique of Tests of Significance in Survey Research." *American Sociological Review*, 23: 519–527.

Somers, R. H. 1962. "A New Asymmetric Measure of Association for Ordinal Variables." *American Sociological Review*, 27: 799–811.

The American National Election Studies (ANES). 2008. American National Election Study, 2008: Pre- and Post-Election Survey. ICPSR25383-v2. Ann Arbor, MI: Inter-university Consortium for Political and Social Research [distributor], 2012-08-30.

United States Department of Health and Human Services. 2011. Substance Abuse and Mental Health Services Administration. Center for Behavioral Health Statistics and Quality. National Survey on Drug Use and Health, 2011. ICPSR34481-v2. Ann Arbor, MI: Inter-university Consortium for Political and Social Research [distributor], 2013-06-20.

Upton, Graham J. G. 1978. *The Analysis of Cross-Tabulated Data*. New York: John Wiley & Sons, Inc.

Van Belle, Gerald. 2002. *Statistical Rules of Thumb*. New York: John Wiley & Sons, Inc.

Yates, F. 1934. "Contingency Tables Involving Small Numbers and the χ^2 Test." *Journal of the Royal Statistical Society*, Supplement 1: 217–235.

Yule, George Udny. 1900. "On the Association of Attributes in Statistics: With Illustrations from the Material of the Childhood Society, &c." *Philosophical Transactions of the Royal Society of London*, Series A, 194: 257–319.

Yule, George Udny. 1912. "On the Methods of Measuring the Association Between Two Variables." *Journal of the Royal Statistical Society*, 75: 576–642.

Zhang, Jun, and Kal F. Yu. 1998. "What's the Relative Risk?" *Journal of the American Medical Association*, 280: 1690–1691. DOI:10.1001/jama.280.19.1690.

3

Contingency Tables in Three Dimensions

Chapter 2 addressed univariate goodness of fit as well as independence and association in two-dimensional contingency tables. This chapter extends the discussion of contingency tables to three dimensions, moving from tables that contain an X and a Y variable to those that contain X, Y, and Z measures. As the chapter explains, inferential tests such as the Cochran-Mantel-Haenszel (Cochran 1954, Mantel and Haenszel 1959) and Breslow-Day (Breslow and Day 1980) procedures examine relationships between two variables given the presence of a third measure. While statistics texts vary in their treatment of these tests – some authors prefer to move directly to log-linear models after covering two-dimensional contingency tables – the current text considers tests designed for three-dimensional tables containing nominal measures useful for communication research. In this regard, the chapter follows scholarship in health communication (Heuer, Maclure, and Puhl 2011, Jarlenski and Barry 2013) as well as public health more generally (Kuritz, Landis, and Koch 1988, Landis, Heyman, and Koch 1978, Preisser and Koch 1997). Demonstrating how the tests can be applied in content analyses and survey research, the chapter includes examples from the horse-racing study, the 2011 National Survey on Drug Use and Health, and the 2008 American National Election Study.

Moving from Two to Three Dimensions

Before discussing the Cochran-Mantel-Haenszel and Breslow-Day procedures, the chapter covers basic terminology in the multivariate analysis of categorical data. To facilitate the discussion, the chapter draws on data from the horse-racing study (Denham 2014). While previous examples referred to equine drug use, the

Table 3.1   Cross-tabulations of time period by horse injury or death mentions in two newspapers

| Newspapers | Time Frame | Horse Injury or Horse Death Mentions | | |
		Mention	*No Mention*	*Totals*
The New York Times	Before First Report	*a* 38 (21.5%)	*b* 139 (78.5%)	177
	After First Report	*c* 52 (36.4%)	*d* 91 (63.6%)	143
	Totals	90	230	320
Los Angeles Times	Before First Report	13 (17.8%)	60 (82.2%)	73
	After First Report	32 (33.3%)	64 (66.7%)	96
	Totals	45	124	169
Totals	Before First Report	51 (20.4%)	199 (79.6%)	250
	After First Report	84 (35.1%)	155 (64.9%)	239
	Totals	135	354	489

focus here involves mentions of horse injuries or horse deaths in news copy. Table 3.1 contains frequency data showing horse injury or horse death mentions in *The New York Times* and the *Los Angeles Times* across two time periods. Because *The New York Times* focused on injured and deceased horses in its investigative series, the content analysis anticipated increases in references to injuries/fatalities in other news outlets following the initial report in *The New York Times*.

Regarding statistical terminology, *partial tables* display data to be tested for *conditional association*, which refers to the presence of at least one significant relationship between two variables at a fixed level of a third. In Table 3.1, one would observe a conditional association if a significant relationship emerged between mentions of horse injuries/horse deaths and time period in *The New York Times* or the *Los Angeles Times*. If a relationship did not appear in either outlet, then one would conclude *conditional independence*, indicating that odds ratios in each partial table did not differ (statistically) from 1.0.

In a three-dimensional contingency table, a *marginal table* contains cross-classified data summed across a third measure. In Table 3.1, the marginal table contains 489 observations, reflecting 320 reports in *The New York Times* and 169 in the *Los Angeles Times*. A *marginal association* occurs when a bivariate relationship emerges in the marginal table. When a relationship is not observed, one concludes *marginal independence*. Notably, in a three-way table, marginal independence does not imply conditional independence, and vice versa. In fact, *Simpson's paradox* (Simpson 1951) may occur when a marginal association moves in the opposite direction of a conditional relationship (Azen and Walker 2011, 86). Without due consideration to descriptive statistics at each level of a contingency table, researchers may draw conclusions based on an *ecological fallacy*, which occurs when the characteristics of individuals are misinterpreted based on aggregate data (see Blyth 1972).

Analyses of three-dimensional contingency tables also may involve testing for *homogenous association*, which occurs when odds ratios in partial tables *do not* differ from one another. In Table 3.1, homogenous association would be present if (a) relationships showed conditional independence, or (b) significant relationships appeared (comparatively) equal. As Azen and Walker (2011, 87) pointed out, while conditional independence implies homogenous association, homogenous association does not imply conditional independence; that is, statistically significant associations observed in partial tables may not differ from one another. The chapter now addresses the first of two inferential tests for analyzing three-way contingency tables.

Cochran-Mantel-Haenszel Test

The Cochran-Mantel-Haenszel (C-M-H) procedure (Cochran 1954, Mantel and Haenszel 1959) tests odds ratios for conditional independence. From a methodological standpoint, the literature addressing the C-M-H procedure is well-developed, in part because statisticians proposed the test more than five decades ago, but also because, unlike the Breslow-Day (B-D) test, reviewed later in the chapter, the C-M-H procedure is not limited to $2 \times 2 \times k$ tables; additional categories can be added if necessary (Kuritz, Landis, and Koch 1988, Mantel 1963; see also Agresti 1990, 230–235, Bishop, Fienberg, and Holland 1975, 146–147). This makes the test applicable to categorical data analyses containing polytomous measures.

In communication contexts, health scholars have used the C-M-H test to analyze media use (Dennison, Erb, and Jenkins 2002, Tucker and Bagwell 1991, Tucker and Friedman 1989), media effects (Chen et al. 2002, Eriksson, Maclure, and Kragstrup 2005, Schade and McCombs 2005), and media content (Jarlenski and Barry 2013). Among these studies, Dennison, Erb, and Jenkins (2002) analyzed the television and video viewing habits of a multiethnic, low-income, preschool population. The study examined associations between time spent watching television and videos, the presence of a TV in the bedroom, and the prevalence of overweight children. In a content analysis, Jarlenski and Barry (2013) studied news media coverage of trans fats, using the C-M-H procedure to examine trends in news representations across three time periods. Among their findings, the authors observed that stories mentioning risks for heart disease appeared more likely to mention governmental actions than articles that did not mention heart problems.

Examining the data in Table 3.1, articles published in *The New York Times* following the first investigative report contained more references to horse injuries/horse deaths than articles published prior to the first report, and a similar pattern emerged in the *Los Angeles Times*. In *The New York Times*, the odds of an article published in the first period containing an injury or fatality reference were .478 times the odds of an article in the second period containing such a reference. In the *Los Angeles Times*, the odds ratio was similar, at .433, and the odds ratio for the marginal table was .473. Given those measures of association (and the presence of three categorical variables), the chapter first uses the C-M-H procedure to test the null hypothesis of conditional independence. The formula for the C-M-H test follows a large-sample chi-squared distribution with 1 degree of freedom (see Cochran 1954, Mantel 1963, Mantel and Haenszel 1959) and takes the following form:

$$\frac{\left[\sum_k \left(n_{11k} - \mu_{11k}\right)\right]^2}{\sum_k Var\left(n_{11k}\right)}$$

The statistic reflects the summation of observed minus expected scores, their quantities squared, and then divided by the variance. Using notation from Agresti (2007), where row totals are represented as $(n_{1+k},\ n_{2+k})$ and column totals are represented as $(n_{+1k},\ n_{+2k})$, expected values and the variance can be calculated using the following formulas:

$$\mu_{11k} = E\left(n_{11k}\right) = \frac{n_{1+k}n_{+1k}}{n_{++k}}$$

$$Var\left(n_{11k}\right) = \frac{n_{1+k}n_{2+k}n_{+1k}n_{+2k}}{n^2_{++k}\left(n_{++k} - 1\right)}$$

where μ_{11k} represents the mean (ultimately an expected value) and $Var(n_{11k})$ represents the variance for the cells formed at the intersection of the first row and first column in each partial table. The C-M-H test cannot be calculated without these preliminary formulas, and thus the first step in calculating the C-M-H statistic is to establish an expected value for the first cell in each partial table. Drawing on the formula shown above, expected values for *The New York Times* and the *Los Angeles Times* can be respectively calculated as follows:

$$\mu_{111} = E\left(n_{111}\right) = \frac{n_{1+1}n_{+11}}{n_{++1}} = \frac{177(90)}{320} = 49.78$$

$$\mu_{112} = E\left(n_{112}\right) = \frac{n_{1+2}n_{+12}}{n_{++2}} = \frac{45(73)}{169} = 19.44$$

The next step is to calculate the respective differences between observed and expected cell counts:

$$n_{111} - \mu_{111} = 38 - 49.78 = -11.78$$

$$n_{112} - \mu_{112} = 13 - 19.44 = -6.44$$

Calculations of differences between observed and expected counts are followed by calculations of the variances for the respective cells:

$$Var\left(n_{111}\right) = \frac{n_{1+1}n_{2+1}n_{+11}n_{+21}}{n_{++1}^2\left(n_{++1}-1\right)} = \frac{177\left(143\right)90\left(230\right)}{320^2\left(320-1\right)} = \frac{523,937,700}{32,665,600} = 16.04$$

$$Var\left(n_{112}\right) = \frac{n_{1+2}n_{2+2}n_{+12}n_{+22}}{n_{++2}^2\left(n_{++2}-1\right)} = \frac{73\left(96\right)45\left(124\right)}{169^2\left(169-1\right)} = \frac{39,104,640}{4,798,248} = 8.15$$

With all values for the C-M-H formula determined, the test statistic can then be calculated:

$$\text{C-M-H} = \frac{\left[\sum_k\left(n_{11k}-\mu_{11k}\right)\right]^2}{\sum_k Var\left(n_{11k}\right)} = \frac{\left[-11.78-6.44\right]^2}{12.04+8.15} = \frac{332.00}{24.19} = 13.72$$

Recalling that a C-M-H test follows a chi-square distribution with 1 degree of freedom, a researcher would note that 13.72 exceeds the chi-square critical value of 3.84 at the alpha level of .05, thus allowing the researcher to conclude a conditional association between time period and mentions of horse injuries or horse fatalities. This means that at least one of the partial tables contains an odds ratio significantly different from 1.0. One can then use the Breslow-Day test to assist with interpretation and to indicate whether interpretation of a common odds ratio is appropriate.

Breslow-Day Test

The Breslow-Day procedure (Breslow and Day 1980) tests the null hypothesis that relationships between variables X and Y are (statistically) equal at each level of Z; an outcome of $p < .05$ indicates differences in association and a rejection of the null hypothesis. As examples of research applications, in a study examining the characteristics of adolescent chat-room users, Beebe et al. (2004) compared behavioral odds of users with nonusers, employing the B-D test to draw statistical inferences. In examining online news portrayals of obesity, Heuer, McClure, and Puhl (2011) used the procedure to conclude that, relative to non-overweight individuals, obese individuals were more likely to have their heads removed from photos, to be portrayed showing only their lower bodies, and to be shown eating or drinking.

Use of the Breslow-Day test is limited to $2 \times 2 \times k$ tables, and it therefore may not prove sufficient for answering more complex research questions in analyses containing three categorical variables. Additionally, the formula for calculating the B-D test is somewhat complex; however, statistical software packages such as SPSS will include B-D results along with C-M-H output, if requested. In a software analysis for the current example, the B-D procedure showed homogenous association between time frame and mentions of horse injuries or horse deaths in *The New York Times* and the *Los Angeles Times* χ^2 (1, 489) = .048, p = .827. Coupled with the C-M-H findings, these results suggest that both of the partial odds ratios were different than 1.0 and thereby justify the calculation of a common odds ratio for the three-way table.

The common odds ratio, termed the *Mantel-Haenszel estimate* (see Azen and Walker 2011, 88), can be calculated using the following formula:

$$\frac{\sum_{k}\left(n_{11k}n_{22k} / n_{++k}\right)}{\sum_{k}\left(n_{12k}n_{21k} / n_{++k}\right)}$$

Examining this equation, one observes elements of the cross-product ratio illustrated in Chapter 2. In this case, for each partial table, the numerator consists of $(a)(d)/n$ and the denominator consists of $(b)(c)/n$, with the Mantel-Haenszel estimate reflecting the summation of these quantities. Applied to the data in Table 3.1, the common odds ratio would be calculated in the following manner:

$$\frac{38(91) / 320 + 13(64) / 169}{139(52) / 320 + 60(32) / 169} = \frac{10.81 + 4.92}{22.59 + 11.36} = .463$$

With homogenous association satisfied by the Breslow-Day test, one would conclude that, on average, the odds of an article published in period one containing a reference to a horse injury or horse death were .463 times the odds of an article published in period two containing such a reference. Statistical software shows this association significant at $p < .001$, but a researcher can also calculate confidence intervals for each of the odds ratios. If the lower and upper bounds of a given confidence interval include the value 1.0, then the appropriate conclusion is independence (i.e., a failure to reject the null hypothesis). Conversely, if a confidence interval does not include 1.0, then an association may exist.

The calculations involved in creating confidence intervals provide a glimpse of *log odds ratios*, which techniques such as logistic regression analysis produce as parameter estimates. Given the centrality of log odds to categorical statistics, in general, it is important for readers to appreciate the fundamental properties of logarithms. In mathematics, the logarithm (or log) of a number is an exponent to which a fixed value (i.e., a base) must be raised to (re)produce the original number. For example, the logarithm of the number 100 to base 10 is 2, such that $100 = 10 \times 10 = 10^2$. Miller (2012, 1) expressed the relationship between logs and exponents in the following manner:

$$\log_b a = x \Leftrightarrow a = b^x$$

"The log of *a*, base *b*, equals *x*," Miller (2012, 1) explained, "which implies that *a* equals *b* to the *x* and vice versa." Through a series of calculations, the author noted, one can also show the following:

$$\log_b\left(b^x\right) = x$$

The logarithm of b^x cancels the exponentiation, resulting in *x*. For purposes of categorical statistics – and more specifically, for purposes of moving from multiplicative equations to additive linear expressions – Miller (2012, 1) noted that "An important property of logarithms is that the logarithm of the product of two variables is equal to the sum of the logarithms of those two variables… Similarly, the logarithm of the ratio of two variables is equal to the difference in their logarithms." Thus, for variables *X* and *Y*:

$$\log_b\left(XY\right) = \log_b X + \log_b Y$$

and

$$\log_b\left(\frac{X}{Y}\right) = \log_b X - \log_b Y$$

Given the focus on odds ratios, it is important to note that the log of 1 is 0 and that log odds have a theoretical range of minus infinity to plus infinity (see Nussbaum 2015, 263). Because an odds ratio cannot be expressed as a negative number, its distribution cannot show symmetry around the value 1 (see Agresti 2007, 30–31, Azen and Walker 2011, 52). Statisticians therefore transform odds ratios to log odds ratios, where initial odds values of less than 1 are expressed as negative numbers and values greater than 1 are expressed as positive numbers.

While the distribution for an odds ratio cannot be estimated using a standard normal distribution, the distribution of a log odds ratio can, facilitating the calculation of a confidence interval. To that end, the log odds ratio can be expressed as $\ln(\text{OR}) \pm z * \text{SE}_{\ln(\theta)}$. In the equation, z represents a number from the standard normal table corresponding to, in this case, a 95% confidence interval.

To begin, the standard error for a log odds ratio is calculated based on the following formula:

$$SE_{\ln OR} = \sqrt{\frac{1}{n_{11}} + \frac{1}{n_{12}} + \frac{1}{n_{21}} + \frac{1}{n_{22}}}$$

where n_{11}, n_{12}, n_{21}, and n_{22} respectively correspond to cells *a*, *b*, *c*, and *d* in the first 2 × 2 partial table displayed in Table 3.1. Standard errors for log odds ratios of *The New York Times* and the *Los Angeles Times* would thus be calculated in the following manner:

$$\text{NYT} = \sqrt{\frac{1}{38} + \frac{1}{139} + \frac{1}{52} + \frac{1}{91}} = \sqrt{.026 + .007 + .019 + .010} = .249$$

$$\text{LAT} = \sqrt{\frac{1}{13} + \frac{1}{60} + \frac{1}{32} + \frac{1}{64}} = \sqrt{.077 + .017 + .031 + .016} = .375$$

After calculating standard errors for the log odds of *The New York Times* and the *Los Angeles Times*, the next step is to use those values in calculating 95% confidence intervals. The value of z, from the standard normal distribution, is 1.96, and as part of the calculations, it must be multiplied by both standard errors, and then added to either side of the logs of the original odds ratios for the two newspapers. Once those steps have been taken, the estimates can be exponentiated back to the level of odds ratios.

$$\text{NYT} = \ln(.478) \pm 1.96(.249) = -.738 \pm .488 = [-1.226, -.25]$$

$$\text{NYT} = [e^{-1.226}, e^{-.25]} = [.293, .779]$$

$$\text{LAT} = \ln(.433) \pm 1.96(.375) = -.837 \pm .735 = [-1.572, -.102]$$

$$\text{LAT} = \left[e^{-1.572}, e^{-.102}\right] = [.207, .903]$$

Based on these calculations, a researcher would conclude with 95% certainty that the "true" odds ratio for *The New York Times* lies between .293 and .779, and the odds ratio for the *Los Angeles Times* lies between .207 and .903. Neither confidence interval includes the value 1.0, and thus the odds ratios for both newspapers indicate an association and a rejection of the null hypothesis of independence.

An Example in Public Health

A second example on the use of Cochran-Mantel-Haenszel and Breslow-Day tests comes from data gathered in the 2011 National Survey on Drug Use and Health (United States Department of Health and Human Services 2011). As shown in Table 3.2, instead of a 2 × 2 × 2 analysis, this example involves a 2 × 2 × 6 contingency table, with two categories of sex positioned across whether or not respondents had experimented with marijuana. A six-category race measure serves as a control, or stratifier. The chapter includes this example to illustrate the analysis of a three-dimensional contingency table derived from a large public dataset, and to demonstrate the importance of examining descriptive statistics, odds ratios, and confidence intervals.

Given the data in Table 3.2, a researcher interested in whether race would affect the relationship between sex and experimentation with marijuana would report, first, that a Cochran-Mantel-Haenszel test indicated the presence of a conditional association χ^2 (1, 18,980) = 15.841, $p < .001$. Examining odds ratios and confidence intervals, it appears the strongest association occurred among

Table 3.2 Cross-tabulations of sex and marijuana experimentation with race as control measure

Race	Sex	*Experimented with Marijuana*			OR	*95% CI*	
		Yes	*No*	*Totals*		*Lower Bound*	*Upper Bound*
White	Male	1,048 (18.4%)	4,656 (81.6%)	5,704			
	Female	869 (16.2%)	4,511 (83.8%)	5,380			
	Totals	1,917	9,167	11,084	1.17	1.058	1.290
Black	Male	264 (20.0%)	1,058 (80.0%)	1,322			
	Female	190 (15.0%)	1,080 (85.0%)	1,270			
	Totals	454	2,138	2,592	1.42	1.156	1.741
Native American,	Male	47 (25.3%)	139 (74.7%)	186			
Alaskan, Pacific	Female	49 (29.5%)	117 (70.5%)	166			
Islander	Totals	96	256	352	0.81	0.505	1.291
Asian	Male	24 (8.3%)	265 (91.7%)	289			
	Female	19 (6.3%)	283 (93.7%)	302			
	Totals	43	548	591	1.35	0.722	2.520
More than One Race	Male	95 (22.0%)	337 (78.0%)	432			
	Female	105 (23.5%)	342 (76.5%)	447			
	Totals	200	679	879	0.92	0.670	1.259
Hispanic	Male	352 (19.4%)	1,459 (80.6%)	1,811			
	Female	296 (17.7%)	1,375 (82.3%)	1,671			
	Totals	648	2,834	3,482	1.10	0.944	1.330
Totals	Male	1,830 (18.8%)	7,914 (81.2%)	9,744			
	Female	1,528 (16.5%)	7,708 (83.5%)	9,236			
	Totals	3,358	15,622	18,980	1.17	1.082	1.257

Note: Data were gathered in the 2011 National Survey on Drug Use and Health and made available by the ICPSR.

Black study participants, with 20% of males and 15% of females experimenting with marijuana. A modest association appeared among White respondents, in the same direction, but confidence intervals for all other races and ethnicities included the value 1.0, and therefore a researcher would not report associations among these groups.

In the current analysis, a Breslow-Day test indicated homogenous association χ^2 (5, 18,980) = 5.506, p = .130, justifying the use of a common odds ratio of 1.17. But examining the confidence interval for the common odds ratio (1.082, 1.257), and noting that 72% of the sample consisted of White and Black individuals, a prudent researcher would conclude only a modest association between sex and marijuana experimentation given race as a stratifier. From the standpoint of practical significance, the findings appear limited.

An Example in Political Communication

The third example of a three-dimensional contingency table draws on data gathered in the 2008 American National Election Study (The American National Election Studies 2008) and includes instructions for the Cochran-Mantel-Haenszel and Breslow-Day tests in SPSS. As indicated in Table 3.3, the example involves sex as a predictor of whether respondents ever discussed politics with friends or family members. Race served as a control in the cross-classifications.

Table 3.3 indicates that the odds of White males having discussed politics with friends or family members were 1.68 times the odds of White females having done so. Among Blacks and members of other races, differences between males and females appeared slight. The marginal table indicates that, overall, the odds of males having discussed politics were 1.36 times the odds of females having done so.

As Figure 3.1 illustrates, SPSS can be used to indicate whether differences in odds ratios for Whites, Blacks, and members of other races exceed chance. To set up a three-dimensional contingency table in SPSS, a researcher should do the following:

- Select Analyze > Descriptive Statistics > Crosstabs and enter variables in the appropriate spaces, as shown in Figure 3.1. For the data in Table 3.3, males and females appear in the rows and discussing politics and not discussing politics appear in the columns. The three-level race measure appears in the open space beneath the row and column measures, serving as a control in the 2 × 2 contingency tables.[1]
- Open the Cells window, select Row percentages, and click Continue.
- Open the Statistics window, select Risk as well as Cochran's and Mantel-Haenszel statistics, and click Continue, as shown in Figure 3.1.
- Click OK.

Table 3.3 Cross-tabulations of sex and political discussion with race as control measure

		Discuss Politics with Family and Friends			
Race	*Sex*	*Yes*	*No*	*Totals*	*OR*
White	Male	225 (80.4%)	55 (19.6%)	280	
	Female	251 (70.9%)	103 (29.1%)	354	
	Totals	476	158	634	1.68
Black	Male	91 (75.8%)	29 (24.2%)	120	
	Female	121 (75.2%)	40 (24.8%)	161	
	Totals	212	69	281	1.04
Other Race	Male	31 (72.1%)	12 (27.9%)	43	
	Female	61 (75.3%)	20 (24.7%)	81	
	Totals	92	32	124	0.85
Totals	Male	347 (78.3%)	96 (21.7%)	443	
	Female	433 (72.7%)	163 (27.3%)	596	
	Totals	780	259	1,039	1.36

Note: Data were gathered in the 2008 American National Election Study and made available by the ICPSR.

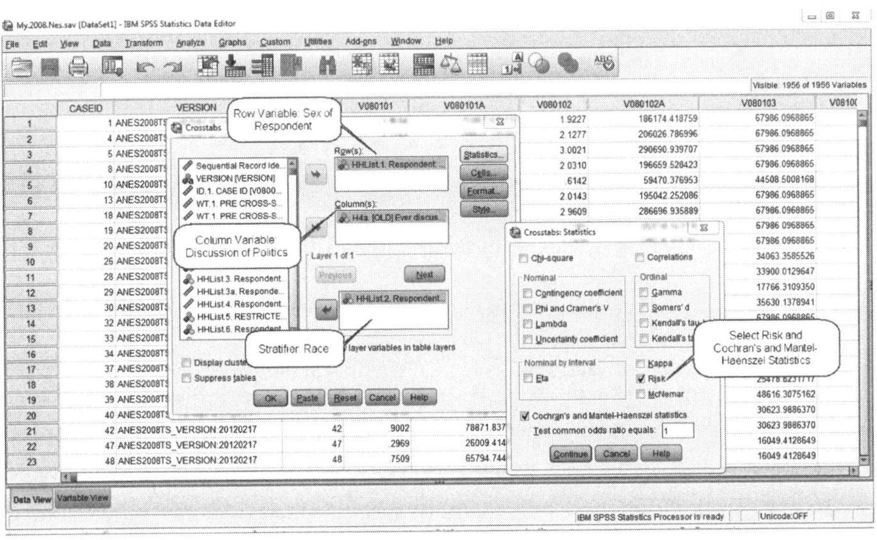

Figure 3.1 SPSS screenshots for Breslow-Day and Cochran-Mantel-Haenszel tests. Source: SPSS® Reprints Courtesy of International Business Machines Corporation, © 2014 International Business Machines Corporation

Table 3.4   Odds ratios reported in Risk function in SPSS

Race		Value	95% Confidence Interval	
			Lower	*Upper*
1. White	Odds ratio for gender: (1 = Male/2 = Female)	1.679	1.155	2.439
	Ever discuss politics with family or friends (1 = Yes)	1.133	1.037	1.238
	Ever discuss politics with family or friends (2 = No)	.675	.506	.900
	N of valid cases	634		
2. Black	Odds ratio for gender: (1 = Male/2 = Female)	1.037	.599	1.798
	Ever discuss politics with family or friends (1 = Yes)	1.009	.882	1.154
	Ever discuss politics with family or friends (2 = No)	.973	.642	1.474
	N of valid cases	281		
3. Other Race	Odds ratio for gender: (1 = Male/2 = Female)	.847	.367	1.954
	Ever discuss politics with family or friends (1 = Yes)	.957	.765	1.198
	Ever discuss politics with family or friends (2 = No)	1.130	.612	2.086
	N of valid cases	124		
Total	Odds ratio for gender: (1 = Male/2 = Female)	1.361	1.019	1.817
	Ever discuss politics with family or friends (1 = Yes)	1.078	1.006	1.156
	Ever discuss politics with family or friends (2 = No)	.792	.636	.987
	N of valid cases	1,039		

Table 3.4 reports odds ratios based on both row and column frequencies. SPSS produces both because a researcher cannot specify which odds ratios he or she is interested in comparing. In this analysis, odds ratios for gender are of interest, and they reflect those calculated through cross-product ratios in Table 3.3. SPSS also produces confidence intervals for odds ratios, as displayed in the table.

As indicated in Table 3.4, the odds of White males discussing politics were approximately 1.68 times the odds of White females doing so. Looking at the lower and upper bounds of the confidence interval for White respondents, one would conclude with 95% certainty that the odds ratio in the broader population would

Table 3.5 Select results of Breslow-Day and Cochran-Mantel-Haenszel tests in SPSS

	Tests of Homogeneity of the Odds Ratio		
	Chi-Squared	*df*	*Sig. (2-sided)*
Breslow-Day	3.400	2	.183
Tarone's	3.400	2	.183

	Tests of Conditional Independence		
	Chi-Squared	*df*	*Sig. (2-sided)*
Cochran's	4.343	1	.037
Mantel-Haenszel	4.034	1	.045

Mantel-Haenszel Common Odds Ratio Estimate			
Estimate			1.357
ln (Estimate)			0.305
Standard Error of ln (Estimate)			0.147
Asymp. Sig. (2-sided)			0.038
95% Confidence Interval	Common Odds Ratio	Lower Bound	1.017
		Upper Bound	1.811
	ln (Common Odds Ratio)	Lower Bound	0.017
		Upper Bound	0.594

fall between 1.155 and 2.439. Also of note are the confidence intervals for races other than Whites, namely their intersections with the value 1.0. As indicated, when 95% confidence intervals include the value 1.0, variables should be considered independent.

Table 3.5 contains select SPSS results for the Breslow-Day and Cochran-Mantel-Haenszel tests. The first section displays the results of the Breslow-Day test for homogenous association. Here the chi-square value of 3.400 with 2 degrees of freedom did not show statistical significance, and thus the null hypothesis could not be rejected; the odds ratios showed homogeneity.

In the second section of Table 3.5, results of the Cochran-Mantel-Haenszel tests appear, with SPSS showing two values. The software contains slight differences in the algorithms used to calculate two very similar tests, but for practical purposes, researchers should use the more conservative of the two, which would be the Mantel-Haenszel test. Because the chi-square value of 4.034 with 1 degree of freedom is statistically significant, one can conclude that at least one of the odds ratios in Table 3.4 is statistically distinct from 1.0. The 95% confidence interval showed an association among White respondents only, and while the Breslow-Day test justifies a common odds ratio (OR = 1.357, $p < .05$), a conservative approach to the results would be to note the comparatively large number of Whites in the analysis, observing their influence in the inferential tests.

Chapter Summary

This chapter has addressed inferential statistics for three-dimensional contingency tables. The chapter focused on calculations associated with the Cochran-Mantel-Haenszel procedure, which tests the null hypothesis of conditional independence, and also reviewed the Breslow-Day test for homogenous association. Examples included data gathered in a content analysis and in survey research, and analyses included nominal data only. Although Mantel (1963) extended the C-M-H statistic to include ordered categories, calculations are somewhat complex and, for the applied statistician, unnecessary given the emergence of log-linear models, discussed in Chapter 4.

Chapter Exercises

1. The following three-dimensional table contains information from the horse-racing study. The table includes measures of time frame, whether newspaper reports mentioned the suspension of an individual associated with horse racing and, as a control variable, the individual newspaper. Use the data to do the following: (a) Calculate an odds ratio for the partial tables as well as the marginal table; (b) calculate an expected cell value and variance based on formulas provided earlier in the chapter; and

| Newspapers | Time Frame | Suspension References in News Reports | | Totals |
		Mention	No Mention	
The New York Times	Before First Report	a 30 (16.9%)	b 147 (83.1%)	177
	After First Report	c 47 (32.9%)	d 96 (67.1%)	143
	Totals	77	243	320
Los Angeles Times	Before First Report	7 (9.6%)	66 (90.4%)	73
	After First Report	16 (16.7%)	80 (83.3%)	96
	Totals	23	146	169
Washington Post	Before First Report	8 (14.8%)	46 (85.2%)	54
	After First Report	16 (30.8%)	36 (69.2%)	52
	Totals	24	82	106
Totals	Before First Report	45 (14.8%)	259 (85.2%)	304
	After First Report	79 (27.1%)	212 (72.9%)	291
	Totals	124	471	595

(c) use the results of those calculations (and the differences between observed and expected frequencies) to calculate the Cochran-Mantel-Haenszel statistic. Given that a Breslow-Day test identified homogenous association, calculate (d) a common odds ratio for the table; and (e) 95% confidence intervals for the odds ratios in the partial tables. Lastly, (f) report on whether the confidence intervals included 1.0; and (g) offer a brief summary of what the statistics mean in terms of substantive relationships between the variables.

2. The following three-dimensional table contains information from the 2008 American National Election Study. The table includes measures of sex, approval/disapproval of the manner in which the president had handled the war in Iraq and, as a control variable, a categorical indicator of political ideology. Use the data to do the following: (a) Calculate an odds ratio for the partial tables as well as the marginal table; (b) calculate an expected cell value and variance based on formulas provided earlier in the chapter; and (c) use the results of those calculations (and the differences between observed and expected frequencies) to calculate the Cochran-Mantel-Haenszel statistic. Given that a Breslow-Day test identified homogenous association, calculate (d) a common odds ratio for the table; and (e) 95% confidence intervals for the odds ratios in the partial tables. Lastly, (f) report on whether the confidence intervals included 1.0; and (g) offer a brief summary of what the statistics mean in terms of substantive relationships between the variables.

| | | *Management of War in Iraq* | | |
Ideology	*Sex*	*Approve*	*Disapprove*	*Totals*
Liberal	Male	*a* 21 (15.8%)	*b* 112 (84.2%)	133
	Female	*c* 15 (8.7%)	*d* 157 (91.3%)	172
	Totals	36	269	305
Conservative	Male	29 (19.2%)	122 (80.8%)	151
	Female	46 (19.9%)	185 (80.1%)	231
	Totals	75	307	382
Moderate	Male	9 (12.7%)	62 (87.3%)	71
	Female	15 (12.1%)	109 (87.9%)	124
	Totals	24	171	195
Totals	Male	59 (16.6%)	296 (83.4%)	355
	Female	76 (14.4%)	451 (85.6%)	527
	Totals	135	747	882

Note

1 SPSS limits the Cochran-Mantel-Haenszel test to $2 \times 2 \times k$ categories, reflecting the statistical limits of the Breslow-Day procedure. Therefore, more advanced analyses in SPSS need to be handled with log-linear models, addressed in Chapter 4.

References

Agresti, Alan. 1990. *Categorical Data Analysis.* New York: John Wiley & Sons.

Agresti, Alan. 2007. *Introduction to Categorical Data Analysis.* Hoboken, NJ: John Wiley & Sons.

Azen, Razia, and Cindy M. Walker. 2011. *Categorical Data Analysis for the Behavioral and Social Sciences.* New York: Routledge.

Beebe, Timothy J., Stephen E. Asche, Patricia A. Harrison, and Kathryn B. Quinlan. 2004. "Heightened Vulnerability and Increased Risk-Taking among Adolescent Chat Room Users: Results from a Statewide School Survey." *Journal of Adolescent Health*, 35: 116–123. DOI:10.1016/j.jadohealth.2003.09.012.

Bishop, Yvonne M. M., Stephen E. Fienberg, and Paul W. Holland. 1975. *Discrete Multivariate Analysis: Theory and Practice.* Cambridge, MA: MIT Press.

Blyth, C. R. 1972. "On Simpson's Paradox and the Sure Thing Principle." *Journal of the American Statistical Association*, 67: 364–366.

Breslow, N. E., and N. E. Day. 1980. *Statistical Methods in Cancer Research: Volume 1 – The Analysis of Case Control Studies.* Lyon: IARC.

Chen, Xinguang, Tess Boley Cruz, Darleen V. Schuster, Jennifer B. Unger, and Carl Anderson Johnson. 2002. "Receptivity to Pro Tobacco Media and Its Impact on Cigarette Smoking among Ethnic Minority Youth in California." *Journal of Health Communication*, 7: 95–111. DOI:10.1080/10810730290087987.

Cochran, William G. 1954. "Some Methods for Strengthening the Common X^2 Tests." *Biometrics*, 10: 417–451.

Denham, Bryan E. 2014. "Intermedia Attribute Agenda-Setting in the New York Times: The Case of Animal Abuse in U.S. Horse Racing." *Journalism & Mass Communication Quarterly*, 91: 17–37. DOI:10.1177/1077699013514415.

Dennison, Barbara A., Tara A. Erb, and Paul L. Jenkins. 2002. "Television Viewing and Television in Bedroom Associated with Overweight Risk among Low-Income Preschool Children." *Pediatrics*, 109: 1028–1035. DOI:10.1542/peds.109.6.1028.

Eriksson, Tina, Malcolm Maclure, and Jakob Kragstrup. 2005. "To What Extent Do Mass Media Health Messages Trigger Patients' Contacts with Their GPs?" *British Journal of General Practice*, 55: 212–217.

Heuer, Chelsea A., Kimberly J. McClure, and Rebecca M. Puhl. 2011. "Obesity Stigma in Online News: A Visual Content Analysis." *Journal of Health Communication*, 16: 359–371. DOI:10.1080/10810730.2010.535108.

Jarlenski, Marian, and Colleen L. Barry. 2013. "News Media Coverage of Trans Fat: Health Risks and Policy Responses." *Health Communication*, 28: 209–216. DOI: 10.1080/10410236.2012.669670.

Kuritz, S. J., J. R. Landis, and G. G. Koch. 1988. "A General Overview of Mantel-Haenszel Methods: Applications and Recent Developments." *Annual Review of Public Health*, 9: 123–160.

Landis, J., Eugene R. Heyman, and Gary G. Koch. 1978. "Average Partial Association in Three-Way Contingency Tables." *International Statistical Review*, 46: 237–254.

Mantel, Nathan. 1963. "Chi-Square Tests with One Degree of Freedom: Extensions of the Mantel-Haenszel Procedure." *Journal of the American Statistical Association*, 58: 690–700.

Mantel, Nathan, and William Haenszel. 1959. "Statistical Aspects of the Analysis of Data from Retrospective Studies of Disease." *Journal of the National Cancer Institute*, 22: 741–748.

Miller, Michael B. 2012. *Mathematics and Statistics for Financial Risk Management.* Hoboken, NJ: John Wiley & Sons.

Nussbaum, E. Michael. 2015. *Categorical and Nonparametric Data Analyses: Choosing the Best Statistical Technique.* New York: Routledge.

Preisser, John S., and Gary G. Koch. 1997. "Categorical Data Analysis in Public Health." *Annual Review of Public Health*, 18: 51–82. DOI: 10.1146/annurev.publhealth.18.1.51.

Schade, Charles P., and Marc McCombs. 2005. "Do Mass Media Affect Medicare Beneficiaries' Use of Diabetes Services? *American Journal of Preventive Medicine*, 29: 51–53. DOI:10.1016/j.amapre.2005.03.003.

Simpson, E. H. 1951. "The Interpretation of Interaction in Contingency Tables." *Journal of the Royal Statistical Society B*, 13: 238–241.

The American National Election Studies. 2008. American National Election Study: ANES Pre- and Post-Election Survey. ICPSR25383-v2. Ann Arbor, MI: Inter-university Consortium for Political and Social Research [distributor], 2012-08-30. DOI:10.3886/ICPSR25383.v2.

Tucker, Larry A., and Marilyn Bagwell. 1991. "Television Viewing and Obesity in Adult Females." *American Journal of Public Health*, 81: 908–911. DOI:10.2105/AJPH.81.7.908.

Tucker, Larry A., and Glenn M. Friedman. 1989. "Television Viewing and Obesity in Adult Males." *American Journal of Public Health*, 79: 516–518. DOI:10.2105/AJPH.79.4.516.

United States Department of Health and Human Services. 2011. Substance Abuse and Mental Health Services Administration. Center for Behavioral Health Statistics and Quality. National Survey on Drug Use and Health, 2011. ICPSR34481-v2. Ann Arbor, MI: Inter-university Consortium for Political and Social Research [distributor], 2013-06-20.

4

Log-linear Analysis

Chapter 3 focused on the Cochran-Mantel-Haenszel and Breslow-Day tests for contingency tables containing three categorical variables. The present chapter moves from inferential *tests* of variable relationships to techniques that *model* associations between multiple categorical measures. In general, models offer greater flexibility than inferential tests do, and the hierarchical structure of a log-linear model, in particular, allows researchers to conceptualize studies in terms of main and interactive effects. In fact, statisticians often characterize log-linear models as analogs of analysis of variance (ANOVA) procedures (see Fienberg 2007, Knoke and Burke 1980), and researchers familiar with ANOVA techniques may notice conceptual similarities between the linear and log-linear approaches.

The current chapter focuses on the general log-linear model, which does not draw distinctions between explanatory and response measures. Instead, the general model treats all variables as outcomes, modeling the natural logs of cell frequencies. Chapter 5 covers logit log-linear analysis, a technique that does draw distinctions between independent and dependent measures. All forms of log-linear analysis belong to a special class of generalized linear models (GLMs) (Nelder and Wedderburn 1972), and the current chapter offers an overview of GLM techniques. The following section reviews the evolution of log-linear modeling and identifies a limited number of analyses in communication research. The chapter then reviews model components and processes using data from the 2008 American National Election Study and the 2012 Monitoring the Future study. As part of this discussion the chapter includes instructions for fitting general log-linear models in SPSS and also addresses visual displays of data.

Categorical Statistics for Communication Research, First Edition. Bryan E. Denham.
© 2017 John Wiley & Sons, Inc. Published 2017 by John Wiley & Sons, Inc.
Companion Website: www.wiley.com/go/denham/categorical_statistics

Development of Log-linear Models

As indicated in Chapter 2, George Udny Yule (1900) viewed categorical variables as inherently categorical, or discrete, and demonstrated how an odds ratio could be calculated through a simple cross-product equation. In 1935, Bartlett used the cross-product ratio in identifying a three-factor interaction in a $2 \times 2 \times 2$ contingency table (Fienberg 2007, 5). While statisticians had long observed *first order* interactions between two variables, Bartlett (1935) produced research showing *second order* interactions among three measures. R. A. Fisher, who statisticians credit with developing maximum likelihood estimation (Fienberg and Rinaldo 2007), consulted with Bartlett, whose work Roy and Kastenbaum (1956) extended to measures containing multiple categories (see also Roy and Mitra 1956).

In the 1960s, statisticians such as Birch (1963) and Mantel (1966) helped to move categorical data analysis closer to the point at which sociologist and statistician Leo Goodman, in a series of papers (1968, 1970, 1971a, 1971b), developed log-linear models for the social sciences.[1] But as important as the Goodman articles were to the development of log-linear modeling, his published studies proved too technical for many researchers (Davis 1978). Statistical software packages had not been developed to any "measureable" degree, and graduate seminars maintained a focus on ANOVA procedures and linear regression models. A dearth of nontechnical material addressed social-science applications of log-linear modeling (Swafford 1980), and only a few studies used the technique (see Knoke 1974).

Accessible studies and monographs based on the work of Goodman, Haberman (1973, 1974a, 1974b), and others began to appear in the early and middle 1980s (Alba 1987, Gilbert 1981, Kennedy 1983), primarily in sociology (Cohen and Cantor 1980, Weil 1982), where researchers analyze social and economic categories and classifications (see Sloane and Morgan 1996).[2] Since that point, researchers in other disciplines, including communication, have used log-linear models to analyze associations among categorical variables. The following section reviews some of those studies.

Examples of Published Research

Relatively few communication scholars used log-linear analysis during its formative period, but as Denham (2002) explained, the differing foci of published studies did reveal the technique's potential to inform communication processes. In a study of political communication, Rust, Bajaj, and Haley (1984) used log-linear analysis to determine which of three media types appeared most efficient for reaching voters during election campaigns, and in the context of crisis communication, Dyer,

Miller, and Boone (1991) applied log-linear analysis to a study of how news services covered the Exxon Valdez oil spill. Scholars also applied the technique in studies of interpersonal and relational communication (Honeycutt et al. 1998, Witteman and Fitzpatrick 1986), nonverbal behavior (Metts and Cupach 1989), group dynamics and conflict (Franz and Jin 1995), deviance (Katz 1994), and community involvement (Rothenbuhler 1991).

More recently, Olekalns and Smith (2000) studied relationships between negotiation strategies and quality of outcomes, Gnisci and Bakeman (2007) analyzed courtroom interactions, while Catellani and Covelli (2013) studied the use of counterfactual communication in politics. Roberts and Liu (2014) used log-linear modeling in a cross-national study of health-related news frames, and their analysis is especially instructive for the current chapter. The authors studied "modal arguments" in editorials across a 35-year period, with arguments consisting of five news frames (economic, security-related, political, welfare-related, cultural) across four reality claims (possible, impossible, inevitable, contingent). The authors created a categorical variable for time, examining five-year increments, and also included a measure indicating the country in which the news reports originated. Four-variable log-linear models then tested relationships among news frames, reality claims, time period, and country of origin. Through their research, Roberts and Liu (2014) demonstrated the potential of log-linear analysis to inform communication processes.

Log-linear Analysis: Fundamentals

The general log-linear model examines associations between two or more categorical variables. Conceptually, the technique can be viewed as a cross-tabulation in more than two dimensions. In log-linear modeling, *categorical factors* may be measured at the nominal or ordinal level (Goodman 1978, 1984), and statistical packages such as SPSS also allow researchers to include *continuous covariates* in analyses (although software systems usually add averages, not unique values, to every cell). Researchers who use log-linear techniques typically analyze whether a *saturated model*, which contains all possible interactions and main effects (and therefore 0 degrees of freedom), can be reduced to a more parsimonious representation of the data. As Powers and Xie (2000, 135) noted, a saturated model "simply parameterizes observed frequencies" and therefore offers little substantive value. In contrast, a *parsimonious model* reproduces cell frequencies in a statistically acceptable manner and increases degrees of freedom by eliminating unnecessary parameters.

From a purely statistical standpoint, then, the best-fitting model is typically one in which the fewest parameters provide an adequate account of cell frequencies (Knoke and Burke 1980, 54). As discussed later in the chapter, researchers generally identify the best-fitting model by comparing likelihood ratio statistics (G^2) across competing representations of the data. Informed by a theoretical framework, scholars often begin with a full hierarchical model and work toward a simpler one. Once a model is selected, its parameter effects can be observed and then exponentiated to form odds ratios.

In log-linear analyses, statisticians use the *multinomial distribution* for predetermined samples and the *Poisson distribution* for modeling an unknown number of counts accumulating across time (Agresti 1990, 37–38). In a practical sense, this means that multinomial distributions should be used in surveys with systematically derived random samples, while Poisson distributions should be used for content analyses that proceed chronologically, without a known sample. (Chapter 10 addresses Poisson regression techniques for the analysis of count data.)

Unlike ANOVA, which uses least squares estimation in parameterization processes, log-linear models use maximum likelihood estimation and the Newton-Raphson algorithm or the iterative proportional fitting algorithm of Deming and Stephan (Haberman 1978, 192). Bishop, Fienberg, and Holland (1975) noted that maximum likelihood estimation is advantageous because it is relatively easy to calculate and, when applied to multinomial data with several zero-count cells, the procedure still produces non-zero estimates. The following section reviews statistical notation in the context of two-way tables, allowing the chapter to then address more complex models.

Two-way Tables

This section of the chapter uses notation from Agresti (1990, 2007), Fienberg (2007), and other scholars who have offered instruction on log-linear analysis. Notationally, a two-way contingency table cross-classifies row variable X by column variable Y, with subscript i signifying rows and j indicating columns. Observed cell frequencies for the ith row and j^{th} column are denoted by f_{ij}. A row total is denoted by $i+$, a column by $+j$, and a grand total by $++$. Thus, n equals f_{++}. When a third variable Z is added, its categories are denoted by subscript k. Symbol π indicates cell probabilities, with μ reflecting expected frequencies, λ indicating parameter effects, and OR indicating odds ratios. The term "log" refers to the natural log of a given value.

For an independence model in a two-way contingency table, row and column marginal totals determine joint cell probabilities (π_{ij}). Thus: $\pi_{ij} = \pi_{i+}\pi_{+j}$. As Agresti (2007, 204–205) noted, cell probabilities serve as parameters for a multinomial distribution, although the general log-linear model uses *expected frequencies* rather than probabilities ($\mu_{ij} = n\pi_{ij}$). Expected frequencies under the Poisson distribution are also denoted by μ_{ij}, and the multiplicative independence model is $\mu_{ij} = \pi_{i+}\pi_{+j}$. Applying the natural log to each side of this equation moves the model from multiplicative to additive, resulting in the following log-linear model of independence:

$$log\,\mu_{ij} = \lambda + \lambda_i^X + \lambda_j^Y$$

where λ is a constant, λ_i^X is a row effect, and λ_j^Y is a column effect (note that X and Y are superscript labels only). For two-way tables, the only remaining model is a saturated one. The saturated model contains the interaction term λ_{ij}^{XY} and indicates that variables X and Y are not statistically independent, but dependent (Agresti 2007). The model is expressed in the following manner:

$$log\,\mu_{ij} = \lambda + \lambda_i^X + \lambda_j^Y + \lambda_{ij}^{XY}$$

As the following section demonstrates, moving from a model containing two variables to one containing three measures results in a marked increase in potential interactions.

Three-way Models

In analyses containing three categorical variables, four types of models exist: (a) mutual independence, (b) joint independence, (c) conditional independence, and (d) homogenous association (Azen and Walker 2011, 137–154, Tang, He, and Tu 2012, 212–218, Powers and Xie 2000, 136–137). This section of the chapter uses a $2 \times 3 \times 3$ display of data gathered in the 2008 American National Election Study (The American National Election Studies 2008) to illustrate and discuss each of the four models. The three categorical factors include sex (Male/Female), race (White/Black/Other Race), and political leaning (Liberal/Conservative/Moderate).[3] The saturated model for the three measures – sex (S), race (R), and political leaning (P) – is represented as:

$$log\,\mu_{ijk} = \lambda + \lambda_i^S + \lambda_j^R + \lambda_k^P + \lambda_{ij}^{SR} + \lambda_{ik}^{SP} + \lambda_{jk}^{RP} + \lambda_{ijk}^{SRP}$$

The model contains three main effects, three two-way (first order) associations, and one three-way (second order) interaction. It also contains 0 degrees of freedom and produces expected frequencies identical to observed values. The researcher is tasked with fitting a model that contains fewer parameters but generates statistically acceptable cell frequencies. To begin a systematic process of model selection, a researcher might first remove the three-factor interaction term, resulting in the model of *homogenous association*. This model, which appears below, contains each of the two-way interactions from the saturated model, in addition to main effects, and suggests that associations between any two variables are consistent across each level of a third measure.

$$\log\mu_{ijk} = \lambda + \lambda_i^S + \lambda_j^R + \lambda_k^P + \lambda_{ij}^{SR} + \lambda_{ik}^{SP} + \lambda_{jk}^{RP}$$

Next, *conditional independence* models contain two two-way interactions. The three models below indicate conditional independence, with combinations of first order associations varying in each model. The first model indicates no association between race and political leaning given sex as a control. The second model indicates independence between sex and political leaning given race as a control, and the third model suggests no association between sex and race with political leaning controlled.

$$\log\mu_{ijk} = \lambda + \lambda_i^S + \lambda_j^R + \lambda_k^P + \lambda_{ij}^{SR} + \lambda_{ik}^{SP}$$

$$\log\mu_{ijk} = \lambda + \lambda_i^S + \lambda_j^R + \lambda_k^P + \lambda_{ij}^{SR} + \lambda_{jk}^{RP}$$

$$\log\mu_{ijk} = \lambda + \lambda_i^S + \lambda_j^R + \lambda_k^P + \lambda_{ik}^{SP} + \lambda_{jk}^{RP}$$

In the first model of conditional independence, where race and political leaning appear independent given the sex measure, the table can be collapsed along race and political leaning, but not sex. Marginal associations for sex and race as well as sex and political leaning appear statistically equal to partial associations; however, the marginal association for race and political leaning does not. Commenting on conditional independence, Powers and Xie (2000, 137) noted, "This is an important model. It means that the marginal association…may be spurious if one ignores a relevant variable…similar to an omitted-variable bias in linear regression." In other words, control measures can prove integral to both statistical and substantive interpretations.

Joint independence models contain just one two-way interaction. The first of the three models below suggests that associations between sex and race are the same at each level of political leaning. The second model implies that associations between sex and political leaning are the same at each level of race, and the third model indicates that associations between race and political leaning are the same at each level of sex.

$$\log\mu_{ijk} = \lambda + \lambda_i^S + \lambda_j^R + \lambda_k^P + \lambda_{ij}^{SR}$$

$$\log\mu_{ijk} = \lambda + \lambda_i^S + \lambda_j^R + \lambda_k^P + \lambda_{ik}^{SP}$$

$$\log\mu_{ijk} = \lambda + \lambda_i^S + \lambda_j^R + \lambda_k^P + \lambda_{jk}^{RP}$$

Lastly, the *mutual independence* model contains no interactions: Sex, race, and political leaning are independent of one another. This model takes the following form:

$$\log\mu_{ijk} = \lambda + \lambda_i^S + \lambda_j^R + \lambda_k^P$$

Having reviewed four types of three-variable log-linear models, the chapter now addresses goodness of fit and the selection of parsimonious models.

Goodness of Fit and Model Selection

Like many social scientists, communication scholars frequently conduct tests that indicate whether a statistically significant association exists between two measures or whether significant differences exist between two groups. Researchers also test relationships involving more than two variables and more than two groups, but in nearly all cases, scholars report on relationships that appear significant at $p < .05$. Parametric techniques frequently include the T-test, the analysis of variance, the Pearson Product Moment Coefficient of Correlation, and ordinary least squares regression. Scholars typically examine cross-tabulated categorical data with chi-square analysis, rejecting or failing to reject the null hypothesis of independence, again with $p < .05$ as a criterion.

Log-linear analysis requires a different approach, as the task of locating a parsimonious model involves the observation of a *nonsignificant* value for G^2, the likelihood ratio statistic. Log-linear models are grounded in the analysis of observed data, with the best-fitting models producing expected values close to observed frequencies. A significant G^2 value indicates significant differences between observed and expected frequencies, and thus a corresponding model does not fit the data. Knoke and Burke (1980) recommended that a G^2 statistic contain a p-value between .10 and .35, as models with p-values greater than .35 risk over-parameterization (i.e., the models fit the data "too well") and models with p-values less than .10 indicate departures between observed and expected cell frequencies. While few statisticians would reject a model in which $p = .40$, establishing a specific range for G^2 and observing that range helps to keep analyses systematic and operationally sound.

In terms of the likelihood ratio itself, statisticians have used the measure since Wilks (1935) developed it, preferring it to a similar chi-square indicator of fit. As Agresti (1989, 297) noted, "An advantage of the likelihood ratio statistic G^2 is that, unlike the Pearson form of statistic, it cannot increase as the model is made more complex. This feature makes it useful for comparing models." The G^2 statistic is also well-suited for maximum likelihood estimation (Bishop, Fienberg, and Holland 1975).

Table 4.1 contains goodness-of-fit statistics for the four types of log-linear models discussed in the previous section, beginning with homogenous association and moving to conditional, joint, and mutual independence, respectively. Model 1, the homogenous association model, does not fit the data. In this model, which contains three main effects and three interactions, the G^2 value of 9.683 (df = 4) is significant at $p < .05$. This indicates a difference between observed and expected frequencies.

Table 4.1   General log-linear analyses of sex, race, and political leaning

Model	Likelihood Ratio	df	Significance
1. {S}{R}{P}{SR}{SP}{RP}	9.683	4	.046
2. {S}{R}{P}{SP}{RP}	9.945	6	.127
3. {S}{R}{P}{SR}{RP}	12.107	6	.060
4. {S}{R}{P}{SR}{SP}	20.774	8	.008
5. {S}{R}{P}{RP}	12.398	8	.134
6. {S}{R}{P}{SP}	21.065	10	.021
7. {S}{R}{P}{SR}	23.227	10	.010
8. {S}{R}{P}	23.519	12	.024

Models 2, 3, and 4 test conditional independence, with Model 2 offering an acceptable fit. Model 2 suggests that associations between sex and race are the same at each level of political leaning, with Models 3 and 4 indicating, or moving toward, differences between observed and expected frequencies. Models 5, 6, and 7 test joint independence, with Model 5 fitting the data and indicating an association between race and political leaning at each level of sex. Models 6 and 7 do not fit the data and indicate differences between observed and expected frequencies. Likewise, Model 8, the mutual independence model, does not fit the data, also showing differences.

At this point, the researcher has two models to consider. Model 5 contains one less interaction than Model 2 and appears to offer a more parsimonious representation of variable relationships. To assess model parsimony empirically, one can examine differences in G^2 values and associated differences in degrees of freedom. Because the difference in two G^2 values is a G^2 value itself, it can be used to test the significance of potential parameters. In this case, the G^2 difference between Models 2 and 5 is 2.453 with 2 degrees of freedom. The chi-square distribution indicates this difference is not statistically significant, and therefore, from a statistical standpoint, Model 5 would be considered the most parsimonious.

It is important to emphasize here that empirical model fitting can and should be driven conceptually. As with other statistical techniques, log-linear analysis should be theoretically informed and not used for quantitative "fishing expeditions." Post-hoc explanations for variable relationships should be avoided as much as possible, as they tend to contribute little to theory development.

Descriptive Statistics and Residuals for the Fitted Model

Table 4.2 contains descriptive statistics and residual values for Model 5. Moving from left to right, the table identifies three categorical variables along with observed and expected frequencies. Had the table contained observed and expected values

Table 4.2 Descriptive statistics for log-linear model containing sex, race, and political leaning

Sex	Race	Political Leaning	Observed Count	(%)	Expected Count	(%)	Residual	Standard Residual	Adjusted Residual	Deviance
Male	White	Liberal	59	6.4	64.347	7.0	−5.347	−.691	−.945	−3.199
		Conservative	103	11.2	97.112	10.6	5.888	.632	.898	3.482
		Moderate	38	4.1	42.240	4.6	−4.240	−.668	−.892	−2.835
	Black	Liberal	54	5.9	41.450	4.5	12.550	1.995	2.663	5.345
		Conservative	35	3.8	43.424	4.7	−8.424	−1.310	−1.752	−3.885
		Moderate	27	2.9	27.634	3.0	−.634	−.122	−.161	−1.119
	Other	Liberal	21	2.3	18.159	2.0	2.841	.673	.879	2.471
		Conservative	15	1.6	15.396	1.7	−.396	−.102	−.133	−.884
		Moderate	10	1.1	12.238	1.3	−2.238	−.644	−.837	−2.010
Female	White	Liberal	104	11.3	98.653	10.8	5.347	.570	.945	3.313
		Conservative	143	15.6	148.888	16.2	−5.888	−.527	−.898	−3.397
		Moderate	69	7.5	64.760	7.1	4.240	.547	.892	2.958
	Black	Liberal	51	5.6	63.550	6.9	−12.550	−1.632	−2.663	−4.737
		Conservative	75	8.2	66.576	7.3	8.424	1.072	1.752	4.228
		Moderate	43	4.7	42.366	4.6	.634	.100	.161	1.130
	Other	Liberal	25	2.7	27.841	3.0	−2.841	−.547	−.879	−2.320
		Conservative	24	2.6	23.604	2.6	.396	.083	.133	.893
		Moderate	21	2.3	18.762	2.0	2.238	.522	.837	2.175

from the saturated log-linear model, the two columns would contain identical frequencies. A reduced model produced statistically acceptable expected counts, as evidenced by a nonsignificant likelihood ratio statistic ($p = .134$).

Four types of residual values appear in Table 4.2, which is based on SPSS analyses of the 2008 ANES data. The first value, the raw residual, reflects the basic difference in observed and expected frequencies. As an example, 59 White males described themselves as liberal, with Model 5 showing an expected value of 64.347; when one subtracts the expected value from the observed, one arrives at a residual value of –5.347. However, because variable categories can vary widely in frequency counts, researchers should also examine standardized residuals. Standardized residuals, also known as Pearson residuals, can be calculated by dividing the raw residual by an estimate of its standard deviation. In this case, the raw residual –5.347 has a standardized value of –.691. This value does not appear exceptional.

Lawal (2003, 154) suggested researchers examine standardized residuals greater than 2.0 for lack of fit. Examining Table 4.2, no values exceed 2.0, but the standardized values for Black males and Black females identifying themselves as liberal do approach that threshold. For Black males, the expected count is substantially lower than the observed value, but for Black females, the opposite pattern appears. A researcher might note such patterns for future scholarship. A researcher also might note adjusted residuals, which divide standardized values by estimates of their standard errors, as well as deviance statistics, which reflect the signed square root of a given cell contribution to G^2. Table 4.2 shows how differing forms of residuals move in similar patterns.

Parameter Estimation

Table 4.3 contains parameter estimates for Model 5. Recalling that a log-linear analysis models the natural logs of expected frequencies, parameter estimates require exponentiation prior to substantive interpretation. Exponentiated main effects indicate the odds of appearing in a given variable category versus the odds of appearing in a variable reference category. Statistical software packages generally use the last category of a variable as the reference, but in most cases, users can either choose the reference category or reassign category numbers (which are simply labels for nominal measures). In Table 4.3, reference categories include females, members of a race apart from White or Black, and political moderates; as shown in the table, these categories do not contain parameter estimates.

Regarding interpretation, main effects often prove inconsequential in the general log-linear model, as the general model does not contain a substantive dependent variable. Interactions supply more useful information, and the joint independence model fitted to the ANES data included an interaction between

Table 4.3 Parameter estimates for log-linear model containing sex, race, political leaning, and interaction of race and political leaning

Parameter	Estimate	Standard Error	Z	Sig.	95% Confidence Interval Lower Bound	Upper Bound
Constant	2.932					
Male	−.427	.068	−6.325	.000	−.560	−.295
Female						
White	1.239	.204	6.073	.000	.839	1.639
Black	.815	.216	3.775	.000	.392	1.237
Other Race						
Liberal	.395	.232	1.698	ns	−.061	.850
Conservative	.230	.241	.954	ns	−.242	.701
Moderate						
White × Liberal	.026	.264	.100	ns	−.490	.543
White × Conservative	.603	.267	2.258	.024	.080	1.126
White × Moderate						
Black × Liberal	.011	.279	.039	ns	−.536	.558
Black × Conservative	.222	.285	.780	ns	−.336	.781
Black × Moderate						
Other Race × Liberal						
Other Race × Conservative						
Other Race × Moderate						

race and political leaning. As indicted, the joint independence model suggests associations between two variables are equal at each level of a third measure. Here, this implies associations between race and political leaning are equal at each level of sex, meaning the data can be collapsed.

Table 4.4 contains a 3 × 3 cross-tabulation of race × political leaning. Frequencies included in the table inform the exponentiation of parameter estimates from Table 4.3. For example, exponentiations of White × liberal (.026) and White × conservative (.603) equal 1.03 and 1.83, respectively. Recalling that other race and moderate served as reference categories in the log-linear analysis, one can apply cross-product ratios to confirm exponentiated values, as shown here:

$$OR_l = \frac{163 \times 31}{107 \times 46} = 1.03 \quad OR_c = \frac{246 \times 31}{107 \times 39} = 1.83$$

Table 4.4 Cross-tabulation of race and political leaning

| | *Political Leaning* | | | |
Race	*Liberal*	*Conservative*	*Moderate*	*Totals*
White	163 (31.6%)	246 (47.7%)	107 (20.7%)	516
Black	105 (36.8%)	110 (38.6%)	70 (24.6%)	285
Other Race	46 (39.7%)	39 (33.6%)	31 (26.7%)	116
Totals	314	395	208	917

Note: These data were gathered in the 2008 ANES and were made available by the ICPSR.

The first of the two odds ratios (1.03) shows independence between White respondents and members of other races in regard to liberal political leaning. The second odds ratio (1.83) shows a greater likelihood of White respondents indicating conservative, consistent with its statistically significant parameter estimate in Table 4.3. The following two odds ratios apply to Black respondents relative to members of other races:

$$OR_l = \frac{105 \times 31}{70 \times 46} = 1.01 \quad OR_c = \frac{110 \times 31}{70 \times 39} = 1.25$$

As with the odds ratios for White respondents, the first of the two odds ratios (1.01) shows independence between Black respondents and members of other races who indicated liberal relative to moderate. The second odds ratio (1.25) indicates a slightly higher likelihood of Black respondents indicating conservative. Overall, the most apparent differences in Table 4.4 indeed occur in the conservative column, with nearly one in two White respondents indicating conservative compared to fewer than 4 in 10 Black respondents and one in three members of other races.

Generalized Linear Models

As indicated at the beginning of this chapter, log-linear models belong to a special class of Generalized Linear Models (GLMs). In 1972 Nelder and Wedderburn generalized classic linear models to include procedures containing nonnormal dependent variables. In doing so, Agresti (1990, 83) noted, the authors advanced "a unified theory that encompasses important models for continuous and categorical variables." Scholars had long used ordinary least squares (OLS) regression to test the effects of multiple predictors, and Nelder and Wedderburn offered scholars a similar approach for categorical data.

Generalized linear models contain three components: *systematic* and *random* components as well as a *link function*. Independent variables form the systematic component, with researchers "fixing" these measures toward the observation of a random component, a response variable containing an exponential distribution (e.g., binomial, multinomial, Poisson, gamma).

Regarding the link function, which connects the systematic and random components, OLS regression uses an identity link in modeling effects of explanatory measures on a continuous dependent variable, while techniques such as logistic regression use a logit link to transform the dependent variable, such that it can be modeled in a linear fashion. As suggested by their name, log-linear models use a log link function to model expected frequencies (see, for discussion, Dunteman and Ho 2006, Lawal 2003, 27–36, McCullagh and Nelder 1989).

In "linearizing" models for procedures containing nonnormal response variables, Nelder and Wedderburn (1972) moved from least squares estimation, the standard estimation procedure for OLS regression, to maximum likelihood estimation. As Dunteman and Ho (2006, 5) explained, "When the distribution of the dependent variable is nonnormal and its variance is a function of its mean, least squares estimates are no longer equal to maximum likelihood estimates as they are for the normal distribution."

Procedurally, statisticians differ on the extent to which researchers should view techniques such as log-linear modeling through a GLM lens, as Fienberg (2000, 643–644) explained in the context of software:

> It is true that computer programs for GLM often provide convenient and relatively efficient ways of implementing basic estimation and goodness-of-fit assessment. But adopting such a GLM approach leads the researcher to ignore the special features of log-linear models relating to interpretation in terms of cross-product ratios and their generalizations, crucial aspects of estimability and existence associated with patterns of zero cells, and the many innovative representations that flow from the basic results of linking sample schemes.

SPSS fits log-linear and logistic regression models through its GLM procedure and also contains separate procedures for the two techniques. For purposes of communication research, either approach will generate reliable results but, as Fienberg (2007) noted, the procedures designed for specific techniques offer more information.

Ordinal Log-linear Analysis

Up to this point in the chapter, the discussion of log-linear analysis has focused on associations among nominal variables. The general log-linear model often proves the most useful at this level, indicating how unordered categorical variables relate to one another. But log-linear models can also measure association at the ordinal level (see, for discussion, Agresti 1981, 1983, 1989, Clogg 1982a, 1982b, Goodman 1979, 1981, 1984, Haberman 1974a, Ishii-Kuntz 1994), revealing patterns that ordinal tests such as Goodman and Kruskal's gamma, Kendall's tau, and Somers' *d* may not capture. Agresti (1984) reviewed a series of ordinal techniques, and this section of the chapter focuses on the uniform association model as well as an extension of that procedure.

As a brief review, two-dimensional contingency tables containing nominal variables are limited to independence and saturated models. But as Agresti (1984, 76) explained, "If one or both variables are ordinal…simple models exist that are more complex and realistic than the independence model yet are unsaturated." Using notation from Agresti, one can assign known scores $\{u_i\}$ and $\{v_j\}$ to respective rows and columns, assuming $u_1 < u_2 < \ldots < u_r$ and $v_1 < v_2 < \ldots < v_c$. As indicated below, in making provision for these scores, the uniform association model includes an additional parameter (β); however, unlike models containing nominal variables, the ordinal model does not require additional association parameters as variable categories increase. The uniform association model is thus expressed as:

$$\log\mu_{ij} = \lambda + \lambda_i^X + \lambda_j^Y + \beta\left(u_i - \bar{u}\right)\left(v_j - \bar{v}\right)$$

where $\sum_i \lambda_i^X = \sum_j \lambda_j^Y = 0$, and degrees of freedom equal $(r-1)(c-1)-1$ (for the added parameter).

To demonstrate the uniform association model, the chapter draws on data gathered in the 2012 Monitoring the Future study of twelfth-grade students in the United States (Johnston et al. 2012). In this survey, students ($N = 2{,}276$) were asked about the frequency with which they read a newspaper as well as the level of risk they perceived to be associated with taking four to five drinks of alcohol per day. One might expect students exposed to the news at higher levels to estimate greater levels of risk given information about accidents, crimes, and so forth. Response options for newspaper exposure included never, a few times per year, once or twice a month, once a week, and almost daily, while risk estimates included none, slight, moderate, and great, with a skew toward the high end.

Given a predetermined sample and two ordinal variables, a log-linear analysis assuming a multinomial distribution first tested the model of independence, resulting in a poor fit ($G^2 = 20.744$, df $= 12$, $p = .054$). However, a second model, which contained the extra association parameter, fit the data, with a G^2 value of 14.997 (df $= 11$) and a p-value of .18. For comparative purposes, Table 4.5 contains parameter estimates for the independence model, and Table 4.6 contains estimates for the ordinal association model.

As indicated in Table 4.6, the parameter estimate for the ordered relationship between newspaper use and estimations of alcohol risk showed significance, in this case indicating slightly higher risk estimates among those who read the newspaper more frequently. Exponentiating the parameter estimate for the association (.057) results in a constant odds ratio of 1.06, and when one exponentiates the log odds for the 95% confidence interval, it appears the true population score ranges between 1.01 and 1.11. Although the odds ratio does not include the value 1.0, a researcher would want to consider potential differences between statistical and substantive significance.

Table 4.5 Parameter estimates for log-linear model containing newspaper use and alcohol risk

Parameter	Estimate	SE	Z	Sig.	95% Confidence Interval	
					Lower Bound	*Upper Bound*
Constant	4.403					
NP Never	1.870	.101	18.501	.000	1.672	2.068
NP Year	1.748	.102	17.149	.000	1.548	1.948
NP Month	1.357	.105	12.865	.000	1.150	1.564
NP Week	1.107	.109	10.206	.000	.895	1.320
NP Day						
No Risk	−3.129	.120	−25.991	.000	−3.365	−2.893
Slight Risk	−2.464	.088	−27.993	.000	−2.637	−2.292
Moderate Risk	−1.371	.055	−25.026	.000	−1.478	−1.263
Great Risk						

Table 4.6 Parameter estimates for log-linear model containing newspaper use, alcohol risk, and ordinal association parameter

Parameter	Estimate	SE	Z	Sig.	95% Confidence Interval	
					Lower Bound	*Upper Bound*
Constant	3.317					
NP Never	2.697	.365	7.381	.000	1.981	3.413
NP Year	2.371	.284	8.337	.000	1.813	2.928
NP Month	1.774	.207	8.563	.000	1.368	2.180
NP Week	1.317	.141	9.350	.000	1.041	1.593
NP Day						
No Risk	−2.747	.196	−13.984	.000	−3.132	−2.362
Slight Risk	−2.205	.139	−15.898	.000	−2.477	−1.933
Moderate Risk	−1.239	.078	−15.883	.000	−1.391	−1.086
Great Risk						
NP × Risk	.057	.024	2.369	.018	.010	.104

Three Ordinal Measures

In certain research contexts, scholars may be interested in analyzing associations among three ordinal variables, and as a guide to such analyses, this section of the chapter focuses on an extension of the linear-by-linear association model. The "homogenous uniform association model," Agresti (1984: 90) noted, contains just three more parameters than the independence model. Degrees of freedom are calculated as $(rcl - r - c - l - 1)$, and the model can be tested and reduced as necessary. Absent the three-way interaction, it is expressed as:

$$\log\mu_{ijk} = \lambda + \lambda_i^X + \lambda_j^Y + \lambda_k^Z + \beta^{XY}\left(u_i - \bar{u}\right)\left(v_j - \bar{v}\right)$$
$$+ \beta^{XZ}\left(u_i - \bar{u}\right)\left(w_k - \bar{w}\right) + \beta^{YZ}\left(v_j - \bar{v}\right)\left(w_k - \bar{w}\right)$$

As an example of how this model might be applied, a researcher may be interested in studying associations between newspaper use, perceptions of the risks associated with anabolic-steroid use in amateur and professional sports, and participation in school-sponsored athletics. One might expect adolescents exposed to newspaper reports at higher levels to estimate greater risk associated with performance-enhancing substances, given news items on hazardous side effects, failed drug tests, and so forth. One also might expect sports participation to affect this association, as adolescent athletes often communicate among themselves about issues such as drugs in sports. Additionally, athletes read the newspaper to stay abreast of the competition and to follow developments in professional athletics.

Drawing on the 2012 Monitoring the Future data (Johnston et al. 2012), a three-variable ordinal log-linear analysis included the five-level newspaper-use measure from the earlier example, perceptions of steroid risk (no risk, slight risk, moderate risk, great risk), and the extent to which respondents participated in school-sponsored athletics (not at all, slight extent, moderate extent, considerable extent, great extent). Table 4.7 includes a series of models testing newspaper exposure (N), perceived steroid risk (S), and participation in athletics (P).

As Agresti (1984) explained, the homogenous uniform association model (see Model 1 in Table 4.7) contains three more parameters than the independence model (Model 8). Just 3 degrees of freedom separate the most complex from the simplest representation of variable relationships. For purposes of this chapter, the models in Table 4.7 are instructive in that several fit the data, and what

Table 4.7   General log-linear analyses of newspaper use, steroid risk perceptions, and participation in school-sponsored athletics

Model	Likelihood Ratio	df	Sig.
1. {N}{S}{P}{NS}{NP}{SP}	91.558	85	.294
2. {N}{S}{P}{NS}{NP}	93.839	86	.264
3. {N}{S}{P}{NS}{SP}	126.652	86	.003
4. {N}{S}{P}{NP}{SP}	95.453	86	.228
5. {N}{S}{P}{NS}	129.106	87	.002
6. {N}{S}{P}{NP}	98.043	87	.196
7. {N}{S}{P}{SP}	133.868	87	.001
8. {N}{S}{P}	136.543	88	.001

would appear the most parsimonious representation (Model 6) does not necessarily offer the best fit. Generally, in a log-linear analysis, a model containing more parameters needs to offer a significantly better fit than one with fewer parameters, and in this case, Model 2, which contains two interactions, fits the data better than Model 6. The G^2 difference, 4.204 with 1 degree of freedom, is significant at $p < .05$. In other G^2 comparisons, Model 1 did not offer a better fit than Model 2, and Model 4 did not offer a better fit than Models 6 and 2. Inspection of the data reveals that perceptions of steroid risk and sports participation were indeed independent of one another, given newspaper use. Those who participated in athletics to a greater extent tended to read the newspaper more frequently, and those who read the newspaper more frequently estimated marginally higher levels of risk.

Table 4.8 contains parameter estimates from Model 2. Exponentiating the interaction parameter estimates of .043 and .059 results in constant odds ratios of 1.04 and 1.06, neither of which suggests great *substantive* importance. Again, in an applied setting, a researcher would want to discuss both statistical and substantive associations, informing readers of whether quantitative findings offer insight into media use and perceptions of risk.

Table 4.8 Parameter estimates for ordinal log-linear model containing newspaper use, steroid risk, sports participation, and two interactions

					95% Confidence Interval	
Parameter	*Estimate*	*SE*	*Z*	*Sig.*	*Lower Bound*	*Upper Bound*
Constant	1.099					
NP Never	3.082	.329	9.357	.000	2.437	3.728
NP Year	2.682	.259	10.355	.000	2.174	3.189
NP Month	1.968	.193	10.208	.000	1.590	2.346
NP Week	1.398	.136	10.258	.000	1.130	1.665
NP Day						
No Risk	−2.709	.193	−14.068	.000	−3.086	−2.331
Slight Risk	−1.870	.128	−14.590	.000	−2.121	−1.619
Moderate Risk	−.727	.071	−10.285	.000	−.865	−.588
Great Risk						
No Participation	.887	.108	8.230	.000	.676	1.099
Slight Extent	−.927	.114	−8.146	.000	−1.150	−.704
Moderate Extent	−.982	.098	−10.069	.000	−1.173	−.791
Considerable Extent	−.919	.082	−11.220	.000	−1.080	−.759
Great Extent						
NP × Steroid Risk	.043	.022	2.010	.044	.001	.085
NP × Participation	.059	.010	5.888	.000	.039	.079

More Complex Models

Log-linear models can be applied to analyses containing more than three categorical variables, and this section of the chapter uses a $2 \times 3 \times 3 \times 3$ model with a single covariate to demonstrate the processes involved. It also includes instructions for conducting a log-linear analysis in SPSS. Grounded in identity politics, the section includes ANES variables addressing sex, race, personal optimism, and national optimism, with survey respondents indicating attitudes of optimistic, pessimistic, or neither optimistic nor pessimistic. Additionally, a continuous covariate indicates the average number of days per week that respondents talked about politics with family and friends during the 2008 election campaigns. In certain research settings, scholars may have theoretical reasons for anticipating covariation. In this case, scholars of political communication might expect responses about personal and national optimism to vary across race, and possibly sex, with frequency of discussion influencing the relationships.

Figure 4.1 contains a screenshot of log-linear options in SPSS (Analyze > General Loglinear Analysis).[4] In this window, the researcher can choose between Poisson and multinomial distributions; because ANES researchers used a predetermined sample in gathering data, the current analysis uses the multinomial option. As indicated in the figure, the analysis included four categorical factors, including

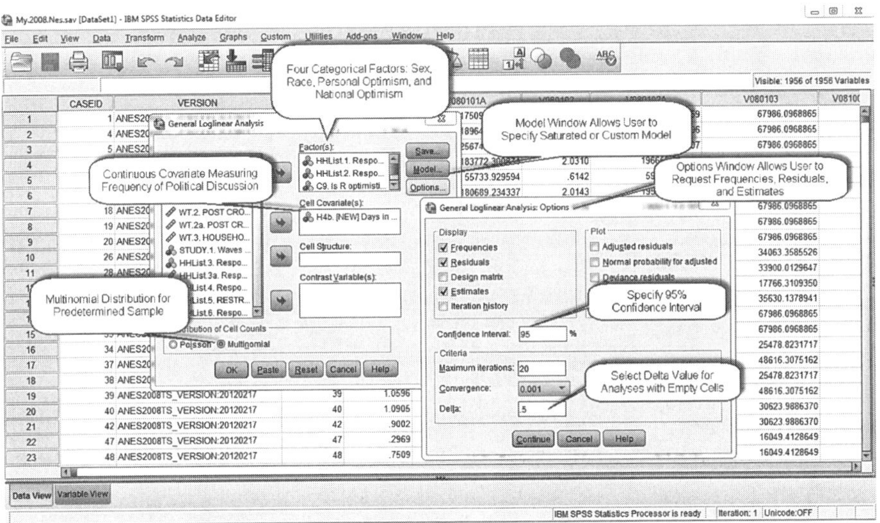

Figure 4.1 SPSS screenshot of general log-linear options. Source: SPSS® Reprints Courtesy of International Business Machines Corporation, © 2014 International Business Machines Corporation

sex (S), race (R), personal optimism (P), and national optimism (N), as well as the covariate measuring frequency of political discussion (D).

Looking to the Options window, one should select, at a minimum, Frequencies, Residuals, and Estimates, while also including a 95% confidence interval as well as a value for delta should zero-count cells appear. Customarily, researchers using log-linear models set delta at .5 (Knoke and Burke 1980), although there is not an absolute standard. Delta should be set to 0 when there are no empty cells. As indicated in the window, plots are also available as part of a log-linear analysis in SPSS, and the default values for maximum iterations and convergence (i.e., 20 and .001) are generally acceptable.

In addition to selecting options shown in Figure 4.1, a researcher also must establish a model to test. In this regard, Figure 4.2 contains the SPSS Model window, and as indicated, one can choose between saturated and custom models. The model constructed in Figure 4.2 includes all terms except a four-way interaction. Designed in a hierarchical manner, the model contains four three-way interactions, six two-way associations, four main effects, and one covariate.[5]

In Table 4.9, Model 1 includes the main effects, interactions, and covariate just described; with a p-value of .217, the model offers a strong fit to the data, yet it includes just 7 degrees of freedom. Proceeding from a perspective in identity politics, a researcher might begin eliminating terms that do not involve race and ethnicity, as reflected in Models 2, 3, and 4. These models examine fit with

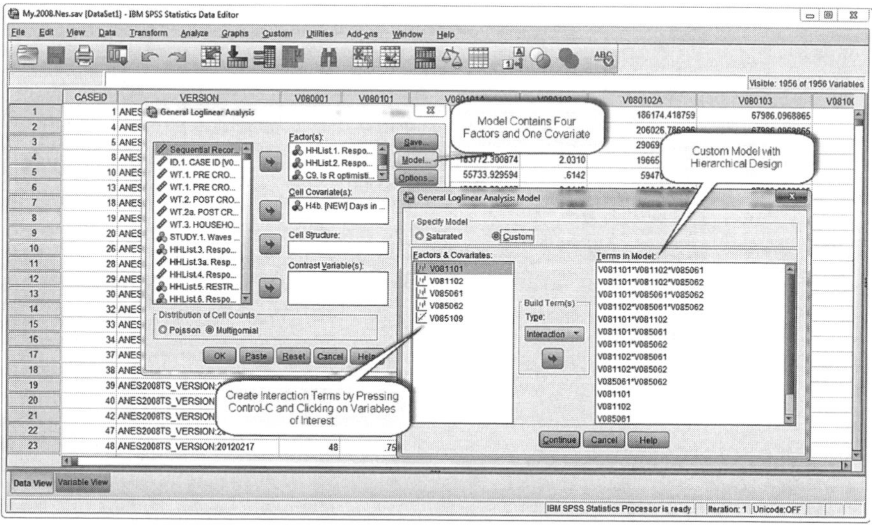

Figure 4.2 SPSS screenshot of log-linear model construction Source: SPSS® Reprints Courtesy of International Business Machines Corporation, © 2014 International Business Machines Corporation

Table 4.9 General log-linear analyses of sex, race, personal optimism, and national optimism with frequency of political discussion as covariate

Model	Likelihood Ratio	df	Sig.
1. {SRP}{SRN}{SPN}{RPN}{SR}{SP}{SN}{RP}{RN}{PN}{S}{R}{P}{N}{D}	9.528	7	.217
2. {SRP}{SRN}{RPN}{SR}{SP}{SN}{RP}{RN}{PN}{S}{R}{P}{N}{D}	11.607	11	.394
3. {SRP}{RPN}{SR}{SP}{SN}{RP}{RN}{PN}{S}{R}{P}{N}{D}	12.891	15	.611
4. {RPN}{SR}{SP}{SN}{RP}{RN}{PN}{S}{R}{P}{N}{D}	14.030	19	.782
5. {SR}{SP}{SN}{RP}{RN}{PN}{S}{R}{P}{N}{D}	27.958	27	.413
6. {SR}{SN}{RP}{RN}{PN}{S}{R}{P}{N}{D}	28.220	29	.506
7. {SR}{RP}{RN}{PN}{S}{R}{P}{N}{D}	29.006	31	.569
8. {SR}{RP}{RN}{S}{R}{P}{N}{D}	307.313	35	.000
9. {SR}{RP}{S}{R}{P}{N}{D}	326.127	39	.000
10. {SR}{S}{R}{P}{N}{D}	336.458	43	.000
11. {S}{R}{P}{N}{D}	348.828	45	.000
12. {PN}{S}{R}{P}{N}{D}	53.857	41	.086
13. {SR}{PN}{S}{R}{P}{N}{D}	48.147	39	.150

three-factor interactions systematically removed. As indicated by the significance levels, none of the resulting models fit the data, although Model 2 appears close. The significance levels for Models 3 and 4 suggest over-parameterization, calling for reduced complexity.

Models 5 through 10 vary in their inclusion of two-factor associations. Continuing with the conceptualized strategy for removing terms, the first three models in the section appear slightly over-parameterized, but a key difference emerges in moving from Model 7 to Model 8. Removing the association between personal and national optimism results in significant differences between observed and expected frequencies. These differences continue through the independence model (Model 11), and the task then becomes one of re-entering theoretically and statistically meaningful associations. In that regard, Model 13 offers a parsimonious explanation of the data and includes just two two-factor interactions in addition to main effects and the communication covariate. While Model 1 contained just 7 degrees of freedom, Model 13 contains 39, and the G^2 difference between the two, 38.619 with 32 degrees of freedom, is not statistically significant; therefore, the simpler model is preferred, provided it fits theoretical expectations.

Table 4.10 contains goodness-of-fit statistics from the SPSS analysis of Model 13. As the table shows, the G^2 value of 48.147 with 39 degrees of freedom

Table **4.10**   SPSS goodness-of-fit display for log-linear model containing sex, race, personal optimism, national optimism, and political discussion covariate

	Goodness-of-Fit Tests[a,b]		
	Value	*df*	*Sig.*
Likelihood Ratio	48.147	39	.150
Pearson Chi-Square	41.566	39	.360

[a] Model: Multinomial
[b] Design: Constant + V081101 + V081102 + V085061 + V085062 + V085109 + V085061 * V085062 + V081101 * V081102

shows an acceptable fit ($p = .15$), and the table also confirms the model design as well as the multinomial distribution. Following that table, Table 4.11a contains descriptive statistics and residuals for male respondents. As indicated earlier, SPSS provides four indicators of differences between observed and expected frequencies, and in this case, no standardized residuals reached 2.0. Examining Table 4.11b, which contains descriptive statistics for female respondents, outlying values did not reach 2.0, but like Black males, Black females appeared more optimistic than (statistically) expected.

Table 4.12 contains parameter estimates for Model 13. Main effects essentially reproduced basic descriptive statistics, indicating, for example, that males were (exp).111 = 1.11 times as likely as females to appear, while Whites were (exp)1.840 = 6.3 times as likely as individuals from other races to appear. Regarding interactions, the referent category for both optimism measures was the third option, in which respondents indicated neither optimism nor pessimism. Parameter estimates show that individuals who indicated personal optimism were (exp)2.078 = 8.0 times as likely to indicate national optimism, and they were (exp).882 = 2.41 times as likely to indicate national pessimism. Those who appeared personally pessimistic were (exp).312 = 1.37 times as likely to express national optimism, and they were (exp)2.392 = 10.94 times as likely to express national pessimism. Thus, overall, the two variables measuring personal and national optimism tended to move in similar directions, helping to explain cell frequencies in the multidimensional table.

Visual Displays

As log-linear models become increasingly complex, visualizing relationships in a full multidimensional contingency table can become difficult. To facilitate interpretation, researchers sometimes draw on elements of mathematical graph theory.

Table 4.11a Descriptive statistics for log-linear model containing sex (males), race, personal optimism, national optimism, and political discussion covariate

| | | | Males | | | | | | |
			Observed Count	(%)	Expected Count	(%)	Residual	Standard Residual	Adjusted Residual	Deviance
Race	Personal	National								
White	Optimistic	Optimistic	128	12.3	131.491	12.7	-3.491	-.326	-.494	-2.625
		Pessimistic	22	2.1	19.825	1.9	2.175	.493	.602	2.140
		Neither	37	3.6	35.596	3.4	1.404	.240	.302	1.692
	Pessimistic	Optimistic	5	0.5	4.614	0.4	.386	.180	.215	.896
		Pessimistic	21	2.0	19.506	1.9	1.494	.342	.433	1.761
		Neither	7	0.7	5.500	0.5	1.500	.641	.753	1.837
	Neither	Optimistic	23	2.2	23.693	2.3	-.693	-.144	-.176	-1.169
		Pessimistic	15	1.4	11.626	1.1	3.374	.995	1.217	2.765
		Neither	34	3.3	40.147	3.9	-6.147	-.990	-1.231	-3.362
Black	Optimistic	Optimistic	43	4.1	45.507	4.4	-2.507	-.380	-.546	-2.208
		Pessimistic	2	0.2	4.719	0.5	-2.719	-1.255	-1.362	-1.853
		Neither	11	1.1	10.058	1.0	.942	.299	.330	1.404
	Pessimistic	Optimistic	0	0.0	.978	0.1	-.978	-.989	-1.037	.000
		Pessimistic	5	0.5	5.160	0.5	-2.719	-1.255	-1.362	-.561
		Neither	3	0.3	1.726	0.2	1.274	.970	1.023	1.821
	Neither	Optimistic	11	1.1	7.276	0.7	3.724	1.386	1.506	3.016
		Pessimistic	2	0.2	2.816	0.3	-.816	-.487	-.517	-1.170
		Neither	17	1.6	15.760	1.5	1.240	.315	.372	1.604
Other Race	Optimistic	Optimistic	31	3.0	29.901	2.9	1.099	.204	.284	1.496
		Pessimistic	3	0.3	5.081	0.5	-2.081	-.925	-.999	-1.778
		Neither	11	1.1	7.900	0.8	3.100	1.107	1.213	2.699
	Pessimistic	Optimistic	2	0.2	.816	0.1	1.184	1.311	1.360	1.893
		Pessimistic	2	0.2	6.073	0.6	-4.073	-1.658	-1.923	-2.108
		Neither	0	0.0	1.082	0.1	-1.082	-1.041	-1.082	.000
	Neither	Optimistic	4	0.4	4.273	0.4	-.273	-.132	-.143	-.727
		Pessimistic	2	0.2	3.788	0.4	-1.788	-.920	-1.028	-1.598
		Neither	15	1.4	11.086	1.1	3.914	1.182	1.340	3.012

Table 4.11b Descriptive statistics for log-linear model containing sex (females), race, personal optimism, national optimism, and political discussion covariate

			Females							
Race	*Personal*	*National*	*Observed Count*	(%)	*Expected Count*	(%)	*Residual*	*Standard Residual*	*Adjusted Residual*	*Deviance*
White	Optimistic	Optimistic	177	17.0	183.158	17.6	−6.158	−.501	−.821	−3.479
		Pessimistic	31	3.0	23.769	2.3	7.231	1.500	1.913	4.058
		Neither	42	4.0	44.864	4.3	−2.864	−.437	−.576	−2.354
	Pessimistic	Optimistic	8	0.8	6.428	0.6	1.572	.622	.810	1.871
		Pessimistic	28	2.7	21.687	2.1	6.313	1.370	1.748	3.782
		Neither	7	0.7	7.799	0.8	−.799	−.287	−.365	−1.230
	Neither	Optimistic	31	3.0	29.664	2.9	1.336	.249	.320	1.653
		Pessimistic	14	1.3	14.384	1.4	−.384	−.102	−.131	−.870
		Neither	43	4.1	49.247	4.7	−6.247	−.912	−1.229	−3.415
Black	Optimistic	Optimistic	77	7.4	68.533	6.6	8.467	1.058	1.527	4.235
		Pessimistic	6	0.6	11.550	1.1	−5.550	−1.642	−1.924	−2.803
		Neither	11	1.1	14.540	1.4	−3.540	−.935	−1.067	−2.478
	Pessimistic	Optimistic	0	0.0	1.438	0.1	−1.438	−1.200	−1.289	.000
		Pessimistic	5	0.5	6.271	0.6	−1.271	−.509	−.572	−1.505
		Neither	3	0.3	3.720	0.4	−.720	−.374	−.422	−1.136
	Neither	Optimistic	7	0.7	12.388	1.2	−5.388	−1.540	−1.747	−2.827
		Pessimistic	4	0.4	3.949	0.4	.051	.026	.028	.320
		Neither	31	3.0	21.610	2.1	9.390	2.041	2.377	4.730
Other Race	Optimistic	Optimistic	29	2.8	26.410	2.5	2.590	.511	.706	2.330
		Pessimistic	4	0.4	3.056	0.3	.944	.541	.574	1.467
		Neither	8	0.8	7.043	0.7	.957	.362	.397	1.428
	Pessimistic	Optimistic	1	0.1	1.725	0.2	−.725	−.553	−.618	−1.044
		Pessimistic	0	0.0	2.303	0.2	−2.303	−1.519	−1.618	.000
		Neither	1	0.1	1.173	0.1	−.173	−.159	−.166	−.564
	Neither	Optimistic	6	0.6	4.205	0.5	1.295	.598	.642	1.708
		Pessimistic	1	0.1	1.436	0.1	−.436	−.364	−.382	−.857
		Neither	8	0.8	10.149	1.0	−2.149	−.678	−.789	−1.951

Table 4.12 SPSS parameter estimates for log-linear model containing sex, race, personal optimism, national optimism, and political discussion covariate

					95% Confidence Interval	
Parameter	*Estimate*	*SE*	*Z*	*Sig.*	*Lower Bound*	*Upper Bound*
Constant	1.888					
Male	.111	.180	.616	.538	–.241	.463
Female						
White	1.840	.141	13.018	.000	1.563	2.117
Black	.773	.161	4.788	.000	.456	1.089
Other Race						
Personal Optimism	–.365	.133	–2.751	.006	–.626	–1.050
Personal Pessimism	–1.920	.233	–8.223	.000	–2.377	–1.462
Personal Neither						
National Optimism	–.689	.141	–4.876	.000	–.966	–.412
National Pessimism	–1.526	.191	–7.975	.000	–1.901	–1.151
National Neither						
Political discussion	.191	.063	3.043	.002	.068	.314
Personal Optimism × National Optimism	2.078	.174	11.948	.000	1.738	2.419
Personal Optimism × National Pessimism	.882	.239	3.686	.000	.413	1.351
Personal Optimism × National Neither						
Personal Pessimism × National Optimism	.312	.360	.867	.386	–.393	1.017
Personal Pessimism × National Pessimism	2.392	.312	7.670	.000	1.780	3.003
Personal Pessimism × National Neither						
Personal Neither × National Optimism						
Personal Neither × National Pessimism						
Personal Neither × National Neither						
Male × White	–.433	.194	–2.228	.026	–.813	–.052
Male × Black	–.496	.225	–2.204	.028	–.938	–.055
Male × Other Race						
Female × White						
Female × Black						
Female × Other Race						

Applied to log-linear analysis, graph *vertices* refer to the categorical variables under study, and *edges*, or lines, represent first-order interactions (Khamis 2011). Darroch, Lauritzen, and Speed (1980) used an edge connecting two vertices to indicate a direct association and the absence of an edge to indicate either independence or conditional independence. Given their approach, Khamis (2011, 34–36) offered the following examples of three-variable models, addressed earlier in the current chapter:

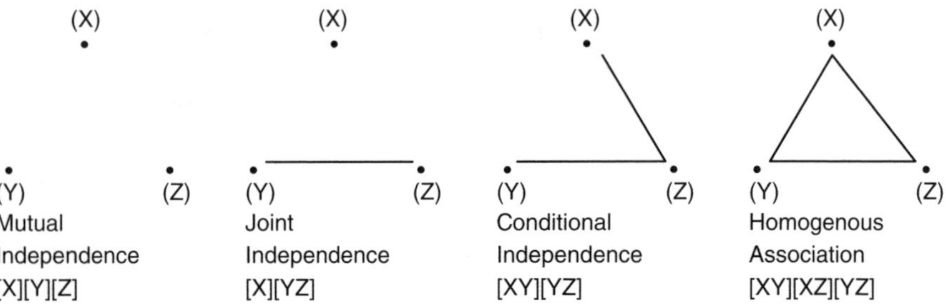

As shown in the figures, edges connecting vertices indicate associations among the respective variables, while the absence of an edge indicates independence or conditional independence. As Darroch, Lauritzen, and Speed (1980, 537) explained, graphs can become especially useful in analyses containing more than three variables, where it may be "difficult a priori to have very precise ideas about the relevant models and where one initially is looking for possible conditional independence among factors." Ideally, a conceptual framework will offer insight on a priori ideas, but in some cases, little is known about associations among the variables in a model.

Khamis (2011, 37–38) and Everitt (1992, 94) illustrated the graphing approach with the following examples involving four categorical variables:

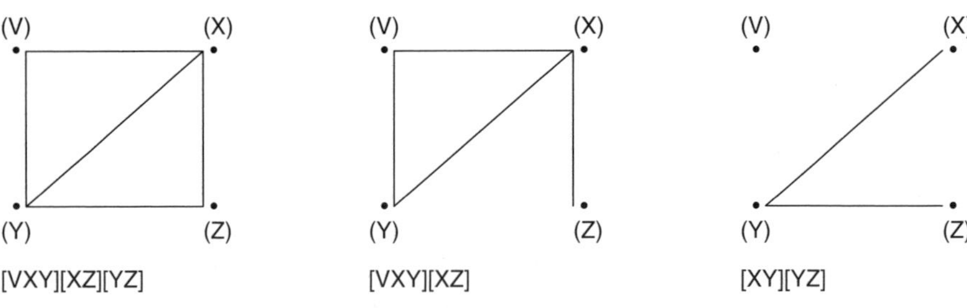

These examples illustrate two- and three-factor associations in four-dimensional contingency tables. In the first case, one observes a single three-factor association along with two two-factor associations. In the second example, the two-factor association YZ drops out, and in the third case, no three-factor associations appear. In this example, variable V is independent of all other measures, whereas variable X is independent of variable Z, contingent on the level of variable Y.

Chapter Summary

This chapter has focused on log-linear analysis as a multivariate statistical technique. General log-linear analysis treats all variables as outcomes and models the natural logs of cell frequencies. The chapter began with a brief history of log-linear modeling and identified a limited number of studies in communication. It reviewed models for two and three nominal measures in addition to models for two and three ordinal variables. The chapter included SPSS instructions in discussing more complex models, in this case those containing four nominal variables and a continuous covariate. It also offered examples of graphical representations.

Chapter Exercises

1. Based on the material covered in this chapter, use the General Loglinear Analysis program in SPSS to analyze relationships among the following categorical variables: sex (Male, Female), race (White, Black, Other Race), and presidential approval (Approve, Disapprove). The first task is to select the appropriate model, beginning with all two-factor interactions and then simplifying. After selecting a model, report its parameter estimates and standard errors in addition to Z values, significance levels, and 95% confidence intervals. Exponentiate the parameter estimates and report the associated odds ratios, offering a brief summary of what the findings suggest about sex, race, and presidential approval.

	Males			*Females*	
Race	*Approval*	*Freq.*	*Race*	*Approval*	*Freq.*
White	Approve	206	White	Approve	227
	Disapprove	408		Disapprove	551
Black	Approve	12	Black	Approve	18
	Disapprove	221		Disapprove	314
Other	Approve	26	Other	Approve	25
	Disapprove	104		Disapprove	125

The data for this exercise, gathered in the 2008 ANES, will need to be entered in grouped frequencies, as shown in the figure below. Data can be entered directly in SPSS or can be entered in an Excel spreadsheet and then imported into SPSS. In either case, categories will need to be assigned numeric labels and weighted (Data > Weight Cases), as shown in the figure. Variables can then be entered in the General Loglinear Analysis program and modeled, as shown in the chapter.

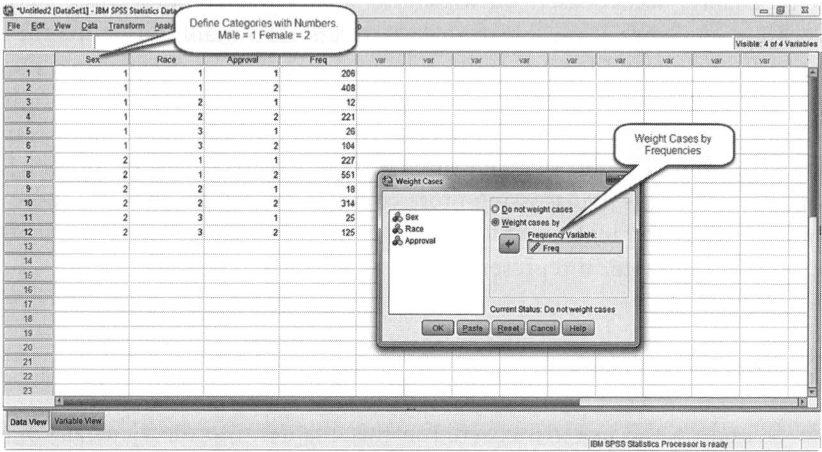

SPSS screenshot of grouped frequencies Source: SPSS® Reprints Courtesy of International Business Machines Corporation, © 2014 International Business Machines Corporation

2. Use the General Loglinear Analysis program in SPSS to analyze relationships among the following categorical variables: read about campaign in newspaper (Yes, No), discussed politics with family or friends (Yes, No), and attitude toward the United States (Optimistic, Pessimistic, Neither Optimistic nor Pessimistic). As with the previous question, data will need to be grouped in SPSS. After weighting, select the appropriate model, beginning with all two-factor interactions and then simplifying. After selecting a model, report its parameter estimates and standard errors in addition to Z values, significance levels, and 95% confidence intervals. Exponentiate the parameter estimates and report the associated odds ratios, offering a brief summary of what the findings suggest about communication and attitudes toward the United States.

Read Newspaper			Did Not Read Newspaper		
Discussed Politics	*Attitude*	*Freq.*	*Discussed Politics*	*Attitude*	*Freq.*
Yes	Optimistic	257	Yes	Optimistic	99
	Pessimistic	61		Pessimistic	25
	Neither	84		Neither	47
No	Optimistic	28	No	Optimistic	30
	Pessimistic	12		Pessimistic	5
	Neither	36		Neither	34

Notes

1 Goodman did not accomplish this task alone, of course, as scholars such as Bishop (1969), Fienberg (1970a, 1970b), and Haberman (1973) made ongoing contributions to the literature. In the physical sciences, work by Grizzle and Williams (1972) also focused on log-linear modeling. Statistics texts on categorical data analysis emerged in the mid-seventies (see Bishop, Fienberg, and Holland 1975, Haberman 1974b, 1978, Plackett 1974), pulling together ideas generated in published research.

2 In statistics, Agresti (1990) added a general text to his previous volume on ordinal data analysis. Additional texts came from Andersen (1990) and Christensen (1990), with a more recent book from Von Eye and Mun (2013).

3 The chapter recognizes that political leaning might be treated as an ordinal variable by moving "moderate" to the center value, but for instructional purposes the variable is retained as a three-level nominal measure.

4 The procedures HILOGLINEAR and LOGLINEAR (syntax only) are also available in SPSS for Windows. Readers should note that HILOGLINEAR and LOGLINEAR differ from GENLOG by producing parameter estimates based on comparisons of observed and expected frequencies. GENLOG uses a reference category.

5 In SPSS, a researcher can create interaction terms by pressing Control-C and clicking on the variables of interest. In the case of ordinal log-linear modeling, an interaction term can be created by clicking on the Transform menu and selecting Compute Variable. For two variables, A and B, an interaction term C can be created by (1) entering variable A in the computation window, (2) clicking on *, (3) entering variable B, and (4) clicking Okay. The interaction term C will appear at the bottom of the variable list. It should be entered as a covariate in the log-linear model containing factors A and B. One can also test models for nominal-ordinal association and ordinal-nominal association. In these models, one can create an ordinal term C by using the Compute Variable option and simply defining ordinal variable B (or ordinal variable A) as C, and then entering C as a covariate with A and B as factors. Doing so takes advantage of the ordinal level.

References

Agresti, Alan. 1981. "Measures of Nominal-Ordinal Association." *Journal of the American Statistical Association*, 76: 524–529. DOI:10.1080/01621459.1981.10477679.

Agresti, Alan. 1983. "A Survey of Strategies for Modeling Cross-Classifications Having Ordinal Variables." *Journal of American Statistical Association*, 78: 184–198. DOI: 10.1080/01621459.1983.10477950.

Agresti, Alan. 1984. *Analysis of Ordinal Categorical Data.* New York: John Wiley & Sons.

Agresti, Alan. 1989. "Tutorial on Modeling Ordered Categorical Response Data." *Psychological Bulletin*, 105: 290–301. DOI:10.1037/0033-2909.105.2.290.

Agresti, Alan. 1990. *Categorical Data Analysis.* New York: John Wiley & Sons.

Agresti, Alan. 2007. *An Introduction to Categorical Data Analysis*, 2nd ed. Hoboken, NJ: John Wiley & Sons.

Alba, Richard D. 1987. "Interpreting the Parameters of Log-Linear Models." *Sociological Methods & Research*, 16: 45–77. DOI:10.1177/0049124187016001003.

Andersen, Erling B. 1990. *The Statistical Analysis of Categorical Data.* New York: Springer-Verlag.

Azen, Razia, and Cindy M. Walker. 2011. *Categorical Data Analysis for the Behavioral and Social Sciences.* New York: Routledge.

Bartlett, M. S. 1935. "Contingency Table Interactions." *Journal of the Royal Statistical Society*, Supplement 2(2): 248–252.

Birch, M. W. 1963. "Maximum Likelihood in Three-Way Contingency Tables," *Journal of the Royal Statistical Society, Series B*, 25: 220–233.

Bishop, Yvonne M. M. 1969. "Full Contingency Tables, Logits, and Split Contingency Tables." *Biometrics*, 25: 383–400.

Bishop, Yvonne M. M., Stephen E. Fienberg, and Paul W. Holland. 1975. *Discrete Multivariate Analysis: Theory and Practice.* Cambridge, MA: MIT Press.

Catellani, Patrizia, and Venusia Covelli. 2013. "The Strategic Use of Counterfactual Communication in Politics." *Journal of Language and Social Psychology*, 32: 480–489. DOI:10.1177/0261927X13495548.

Christensen, Ronald. 1990. *Log-linear Models.* New York: Springer.

Clogg, Clifford C. 1982a. "Some Models for the Analysis of Association in Multiway Cross-Classifications Having Ordered Categories." *Journal of the American Statistical Association*, 77: 803–815. DOI:10.1080/01621459.1982.10477891.

Clogg, Clifford C. 1982b. "Using Association Models in Sociological Research: Some Examples." *American Journal of Sociology*, 88: 114–134.

Cohen, Lawrence E., and David Cantor. 1980. "The Determinants of Larceny: An Empirical and Theoretical Study." *Journal of Research in Crime and Delinquency*, 17: 140–159. DOI:10.1177/002242788001700202.

Darroch, J. N., S. N. Lauritzen, and T. P. Speed. 1980. "Markov Fields and Log-linear Interaction Models for Contingency Tables." *Annals of Statistics*, 8: 539–552.

Davis, James A. 1978. "Hierarchical Models for Significance Tests in Multiway Contingency Tables: An Exegesis of Goodman's Recent Papers." In Leo A. Goodman, *Analyzing Qualitative/Categorical Data: Log-linear Models and Latent-Structure Analysis* (pp. 233–280). Cambridge, MA: Abt Books.

Denham, Bryan E. 2002. "Advanced Categorical Statistics: Issues and Applications in Communication Research." *Journal of Communication*, 52: 162–176. DOI:10.1111/j.1460-2466.2002.tb02537.x.

Dunteman, George H., and Moon-Ho R. Ho. 2006. *An Introduction to Generalized Linear Models*. Thousand Oaks, CA: Sage.

Dyer, Samuel Coad Jr, M. Mark Miller, and Jeff Boone. 1991. "Wire Service Coverage of the Exxon Valdez Crisis." *Public Relations Review*, 17: 27–36.

Everitt, B. S. 1992. *The Analysis of Contingency Tables*, 2nd ed. Boca Raton, FL: CRC Press.

Fienberg, Stephen E. 1970a. "The Analysis of Multidimensional Contingency Tables." *Ecology*, 51: 419–433.

Fienberg, Stephen E. 1970b. "An Iterative Procedure for Estimation in Contingency Tables." *Annals of Mathematical Statistics*, 41: 907–917.

Fienberg, Stephen E. 2000. "Contingency Tables and Log-linear Models: Basic Results and New Developments." *Journal of the American Statistical Association*, 95: 643–647. DOI:10.1080/01621459.2000.10474242.

Fienberg, Stephen E. 2007. *The Analysis of Cross-Classified Categorical Data*, 2nd ed. New York: Springer.

Fienberg, Stephen E., and Alessandro Rinaldo. 2007. "Three Centuries of Categorical Data Analysis: Log-linear Models and Maximum Likelihood Estimation." *Journal of Statistical Planning and Inference*, 137: 3430–3445. DOI:10.1016/j.jspi.2007.03.022.

Franz, Charles R., and K. Gregory Jin. 1995. "The Structure of Group Conflict in a Collaborative Work Group During Information Systems Development." *Journal of Applied Communication Research*, 23: 108–127. DOI:10.1080/00909889509365418.

Gilbert, G. Nigel. 1981. *Modelling Society: An Introduction to Loglinear Analysis for Social Researchers*. London: George Allen & Unwin.

Gnisci, Augusto, and Roger Bakeman. 2007. "Sequential Accommodation of Turn Taking and Turn Length: A Study of Courtroom Interaction." *Journal of Language and Social Psychology*, 26: 234–259. DOI:10.1177/0261927X06303474.

Goodman, Leo A. 1968. "The Analysis of Cross-Classified Data: Independence, Quasi-Independence, and Interactions in Contingency Tables with or Without Missing Entries." *Journal of the American Statistical Association*, 63: 1091–1131. DOI:10.1080/01621459.1968.10480916.

Goodman, Leo A. 1970. "The Multivariate Analysis of Qualitative Data: Interactions among Multiple Classifications." *Journal of the American Statistical Association*, 65: 226–256.

Goodman, Leo A. 1971a. "Partitioning of Chi-Square, Analysis of Marginal Contingency Tables, and Estimation of Expected Frequencies in Multidimensional Contingency Tables." *Journal of the American Statistical Association*, 66: 339–344. DOI:10.1080/01621459.1970.10481076.

Goodman, Leo A. 1971b. "The Analysis of Multidimensional Contingency Tables: Stepwise Procedures and Direct Estimation Methods for Building Models for Multiple Classifications." *Technometrics*, 13: 33–61. DOI:10.1080/00401706.1971.10488753.

Goodman, Leo A. 1978. *Analyzing Qualitative/Categorical Data*. Cambridge, MA: Abt Books.

Goodman, Leo A. 1979. "Simple Models for the Analysis of Association in Cross-Classifications Having Ordered Categories." *Journal of the American Statistical Association*, 74: 537–552. DOI:10.1080/01621459.1979.10481650.

Goodman, Leo A. 1981. "Association Models and Canonical Correlation in the Analysis of Cross-Classifications Having Ordered Categories." *Journal of the American Statistical Association*, 76: 320–334. DOI:10.1080/01621459.1981.10477651.

Goodman, Leo A. 1984. *The Analysis of Cross-Classified Data Having Ordered Categories.* Cambridge, MA: Harvard University Press.

Grizzle, J. E., and O. Williams. 1972. "Log-linear Models and Tests of Independence for Contingency Tables." *Biometrics*, 28: 137–156.

Haberman, Shelby J. 1973. "The Analysis of Residuals in Cross-Classified Tables." *Biometrics*, 29: 205–220.

Haberman, Shelby J. 1974a. "Log-linear Models for Frequency Tables with Ordered Classifications." *Biometrics*, 30: 589–600.

Haberman, Shelby. 1974b. *The Analysis of Frequency Data.* Chicago: University of Chicago Press.

Haberman, Shelby. 1978. *Analysis of Qualitative Data.* New York: Academic Press.

Honeycutt, James M., James G. Cantrill, Pamela Kelly, and David Lambkin. 1998. "How Do I Love Thee? Let Me Consider My Options. Cognition, Verbal Strategies, and the Escalation of Intimacy." *Human Communication Research*, 25: 39–63. DOI:10.1111/j.1468-2958.1998.tb00436.x.

Ishii-Kuntz, Masako. 1994. *Ordinal Log-linear Models.* Thousand Oaks, CA: Sage.

Johnston, Lloyd D., Jerald G. Bachman, Patrick M. O'Malley, and John Schulenberg. 2012. Monitoring the Future: A Continuing Study of American Youth. Funded by National Institute on Drug Abuse. Institutional for Social Research, University of Michigan.

Katz, James E. 1994. "Empirical and Theoretical Dimensions of Obscene Phone Calls to Women in the United States." *Human Communication Research*, 21: 155–182. DOI:10.1111/j.1468-2958.1994.tb00344.x.

Kennedy, John J. 1983. *Analyzing Qualitative Data: Introductory Log-linear Analysis for Behavioral Research.* New York: Praeger.

Khamis, Harry J. 2011. *The Association Graph and the Multigraph for Loglinear Models.* Thousand Oaks, CA: Sage.

Knoke, David. 1974. "Religious Involvement and Political Behavior: A Log-linear Analysis of White Americans, 1952–1968." *The Sociological Quarterly*, 15: 51–65. DOI:10.1111/j.1533-8525.1974.tb02127.x.

Knoke, David, and Peter J. Burke. 1980. *Log-linear Models.* Beverly Hills, CA: Sage.

Lawal, Bayo. 2003. *Categorical Data Analysis with SAS and SPSS Applications.* Mahwah, NJ: Erlbaum.

Mantel, N. 1966. "Models for Complex Contingency Tables and Polychotomous Dosage Response Curves." *Biometrics*, 22: 83–95.

McCullagh, P., and J. A. Nelder. 1989. *Generalized Linear Models*, 2nd ed. London: Chapman and Hall.

Metts, Sandra, and William R. Cupach. 1989. "Situational Influence on the Use of Remedial Strategies in Embarrassing Predicaments." *Communication Monographs*, 56: 151–162. DOI:10.1080/03637758909390256.

Nelder, J. A., and R. W. M. Wedderburn. 1972. "Generalized Linear Models." *Journal of the Royal Statistical Society, Series A*, 135: 370–383.

Olekalns, Mara, and Philip L. Smith. 2000. "Understanding Optimal Outcomes: The Role of Strategy Sequences in Competitive Negotiations." *Human Communication Research*, 26: 527–557. DOI:10.1111/j.1468-2958.2000.tb00768.x.

Plackett, R. L. 1974. *The Analysis of Categorical Data*. London: Griffin.

Powers, Daniel A., and Yu Xie. 2000. *Statistical Methods for Categorical Data Analysis*. San Diego, CA: Academic Press.

Roberts, Carl W., and Hexuan Liu. 2014. "On the Cultural Foundations for Universal Healthcare: Implications from Late 20th-Century U.S. and Canadian Health-Related Discourse." *Journal of Communication*, 64: 764–784. DOI:10.1111/jcom.12079.

Rothenbuhler, Eric W. 1991. "The Process of Community Involvement." *Communication Monographs*, 58: 64–78. DOI: 10.1080/03637759109376214.

Roy, S. N., and M. A. Kastenbaum. 1956. "On the Hypothesis of No 'Interaction' in a Multi-way Contingency Table." *Annals of Mathematical Statistics*, 27: 749–757.

Roy, S. N., and S. K. Mitra. 1956. "An Introduction to Some Non-parametric Generalizations of Analysis of Variance and Multivariate Analysis." *Biometrika*, 43: 361–376.

Rust, Roland T., Mukesh Bajaj, and George Haley. 1984. "Efficient and Inefficient Media for Political Campaign Advertising." *Journal of Advertising*, 13: 45–49. DOI: 10.1080/00913367.1984.10672901.

Sloane, Douglas, and S. Philip Morgan. 1996. "An Introduction to Categorical Data Analysis: A Nontechnical Summary." *Annual Review of Sociology*, 22: 351–375.

Swafford, Michael. 1980. "Three Parametric Techniques for Contingency Table Analysis: A Nontechnical Commentary." *American Sociological Review*, 45: 664–690.

Tang, Wan, Hua He, and Xin M. Tu. 2012. *Applied Categorical and Count Data Analysis*. Boca Raton: FL: CRC Press.

The American National Election Studies. 2008. American National Election Study: ANES Pre- and Post-Election Survey. ICPSR25383-v2. Ann Arbor, MI: Interuniversity Consortium for Political and Social Research [distributor], 2012-08-30. DOI:10.3886/ICPSR25383.v2.

Von Eye, Alexander, and Eun-Young Mun. 2013. *Log-linear Modeling: Concepts, Interpretation, and Application*. Somerset, NJ: John Wiley & Sons.

Weil, Frederick D. 1982. "Tolerance of Free Speech in the United States and West Germany, 1970–1979: An Analysis of Public Opinion Survey Data." *Social Forces*, 60: 973–992. DOI:10.1093/sf/60.4.973.

Wilks, S. S. 1935. "The Likelihood Test of Independence in Contingency Tables. *Annals of Mathematical Statistics*, 6: 190–196.

Witteman, Hal, and Mary Anne Fitzpatrick. 1986. "Compliance-Gaining in Marital Interaction: Power Bases, Processes, and Outcomes." *Communication Monographs*, 53: 130–143. DOI:10.1080/03637758609376132.

Yule, George Udny. 1900. "On the Association of Attributes in Statistics: With Illustrations from the Material of the Childhood Society, &c." *Philosophical Transactions of the Royal Society of London*, Series A, 194: 257–319.

5

Logit Log-linear Analysis

Chapter 4 addressed the general log-linear model, a statistical technique used in examining associations among multiple categorical variables. As discussed in the chapter, the general model does not draw distinctions between independent and dependent measures and yields log expected frequencies as a function of associations and interactions. The current chapter moves from symmetrical to asymmetrical in addressing the logit log-linear model, which Knoke and Burke (1980, 24) characterized as an "analog" of ordinary least squares (OLS) regression. Logit analysis models the log odds of a categorical dependent variable as a function of explanatory measures; that is, it models the log of the odds that a given observation will appear in one category of a response variable relative to another (see Borooah 2002, DeMaris 1992, Liao 1994, Powers and Xie 2000). For communication researchers, the technique may prove useful insofar as the preponderance of multivariate analyses involve predicting the behavior of a dependent variable based on a series of independent measures. As the current chapter demonstrates, the logit model also allows more than one response variable to be modeled as a function of categorical factors and continuous covariates.

The chapter begins with a review of studies that have used logit log-linear analysis, and it then covers the fundamental components of logit models. Many of these components resemble those in the general log-linear model and can be interpreted in line with that technique. Regarding examples, the chapter first draws on data from the 2008 American National Election Study (ANES) in examining a logit model containing one categorical response variable. The chapter then uses data from the 2012 Monitoring the Future (MTF) study to analyze a model containing two response measures. Finally, the chapter draws on

Categorical Statistics for Communication Research, First Edition. Bryan E. Denham.
© 2017 John Wiley & Sons, Inc. Published 2017 by John Wiley & Sons, Inc.
Companion Website: www.wiley.com/go/denham/categorical_statistics

data from the 2011 National Survey on Drug Use and Health (NSDUH) in explaining how to conduct logit log-linear analyses in SPSS.

Examples of Published Research

In a study involving sports broadcasting, Coventry (2004) used logit log-linear analysis to examine organizational roles as a function of gender and race. A nominal response measure contained seven categories: anchor, reporter, studio host, play-by-play announcer, game analyst, studio analyst, and competition-level reporter. Gender and race predicted organizational roles in the media outlets Coventry analyzed, with women and members of racial minorities assigned to lower-level, or less prominent, positions. In an earlier study, Mulder (1981) examined the effects of three categorical (or categorized) measures – sex, education level, and age – on perceptions of media credibility. In the context of health communication, Siminoff, Traino, and Gordon (2011) used a hierarchical log-linear model to assess predictors of consent to tissue donation, and in nonverbal research Patterson and Tubbs (2005) as well as Patterson et al. (2007) examined explanatory effects of pedestrian interactions.

In a slightly different context, Denham (2009) used logit log-linear analysis along with logistic regression techniques in a study addressing risk perceptions of anabolic steroid use. In that study, logistic regression identified significant predictors of a three-level ordinal risk measure; however, in addition to identifying measures that *could* be present in predicting risk attitudes, the research sought to identify the explanatory measures that *needed* to be present in order to explain cell frequencies. The study therefore included logit log-linear analysis in addition to logistic regression. Explanatory factors included sex, race, ease of obtaining the substance, friends who used the substance, newspaper use, and drug spot exposure, while continuous covariates measured sensation seeking, depression, and self-esteem. Denham (2009) found that risk perceptions associated with steroids could be explained with relatively simple log-linear models, as could perceptions of the risks associated with using MDMA (i.e., "ecstasy"). But when the factors and covariates were tested in models involving risk perceptions of marijuana use, it became more difficult to explain cell frequencies in the response measure, likely the result of greater marijuana use among adolescents. As an example, while the depression covariate combined with differing categorical factors to explain risk perceptions associated with steroids and ecstasy, no depression-related models explained perceived risks of marijuana use. Thus, although analyses treated a three-level ordinal variable as nominal, logit models offered a degree of insight absent in logistic regression models. In fact, as explained in the chapter addressing ordinal logistic regression, situations arise when ordinal data *need* to be treated as nominal.

Logit Log-linear Analysis: Fundamental Components

Since the point at which Joseph Berkson (1944) applied the term "logit" to the log of the odds, statisticians have developed models for testing the effects of multiple predictor variables on the behavior of dichotomous, polytomous, and ordinal response measures (see, e.g., Bishop 1969, McFadden 1973; see also Agresti 1990, 91–97, Fienberg 2007, 95–109, Haberman 1978, 292–353). The present chapter considers logit models in the context of log-linear analyses (i.e., analyses of contingency tables), where Goodman (1970) focused much of his work. Goodman and his contemporaries first developed a "modified regression approach" that offered researchers an alternative to using ordinary least squares regression in analyses containing dichotomous response variables. Many quantitative researchers had used OLS procedures in modeling binary dependent variables; Aldrich and Nelson (1984, 30) summarized the methodological problems associated with that approach while recommending logit techniques as alternatives to linear probability models:

> The incorrect assumption of linearity will lead to least squares estimates which (1) have no known distributional properties, (2) are sensitive to the range of the data, (3) may grossly underestimate the magnitude of the true effects, (4) systematically yield probability predictors outside the range of 0 to 1, and (5) get worse as standard statistical practices for improving estimates are employed.

The logit model offers comparably stable parameter estimates and, as DeMaris (1992) explained, produces those estimates while controlling the effects of other predictors. Recalling the discussion of odds from Chapter 2 (and notation from Chapter 4), a logit model describing the effects of a single factor on a dichotomous response variable is expressed as:

$$\log\left(\frac{\pi_{1|i}}{\pi_{2|i}}\right) = \lambda + \lambda_i$$

where the log of the odds of appearing in a given category of the response variable is a function of a constant plus a single parameter. A logit model containing two categorical explanatory factors ($I \times J \times 2$ table) is expressed as:

$$\log\left(\frac{\pi_{1|ij}}{\pi_{2|ij}}\right) = \lambda + \lambda_i^A + \lambda_j^B$$

A model containing three categorical explanatory factors ($I \times J \times K \times 2$ table) is expressed as:

$$\log\left(\frac{\pi_{1|ijk}}{\pi_{2|ijk}}\right) = \lambda + \lambda_i^A + \lambda_j^B + \lambda_k^C$$

In a logit model, the log odds have a theoretical range of minus infinity to plus infinity (DeMaris 1992), and the dependent measure is assumed to follow a multinomial distribution. Parameter estimates are similar to regression coefficients in OLS models (Knoke and Burke 1980, 28), with positive values increasing the odds on a dependent variable and negative values lowering the odds (see also Allison 1999). While main effects in the general log-linear model offer limited information, their presence in the logit model can prove substantively important, as can interactive effects (Huang and Shields 2000).

In an "exegesis" of the Goodman system, Davis (1978) noted the centrality of three terms: *odds ratios*, *effects*, and *models*. Odds ratios, he explained, functioned as operational components of abstract effects, and researchers using the system generally sought the most parsimonious representation of variable relationships. The observations Davis made in 1978 continue to hold true, as the best-fitting model is assumed to be one in which the fewest parameters reproduce response frequencies in a statistically acceptable manner. The Knoke and Burke (1980) suggestion of locating a G^2 p-value between .10 and .35 applies to both general and logit log-linear models (see also Hagle and Mitchell 1992). In the latter, models may include dichotomous as well as polytomous dependent measures.

Commenting on the Goodman system, Magidson (1978) identified four criteria for choosing a model, stressing the importance of both theory and method: (1) Parameters should represent substantively meaningful quantities; (2) parameters should be expressed in units that can be interpreted without difficulty; (3) parameter estimates should lend themselves to appropriate statistical tests; and (4) a model should provide a parsimonious explanation of the data. The following section examines a logit model containing one dependent variable.

Logit Model with One Response Measure

Drawing on data gathered in the 2008 American National Election Study (The American National Election Studies 2008), the first example of a logit log-linear model uses variables analyzed earlier, in a general log-linear analysis. Variables include sex (S), race (R), personal optimism (P), national optimism (N), and a covariate measuring (average) frequency of political discussion (D) among family and friends (previous analyses contained a dichotomous discussion variable). In this analysis, national optimism has been positioned as a three-level response variable (Optimistic, Pessimistic, Neither Optimistic nor Pessimistic), with sex, race, and personal optimism included as explanatory factors.

Table 5.1 contains eight logit models estimated with statistical software. Applying criteria recommended by Knoke and Burke (1980), Model 1 offers an

Table 5.1   Logit log-linear models including sex, race, personal optimism, and frequency of political discussion as explanatory measures of national optimism

	Dependent Variable	Independent Variables	Covariate	G^2	df	Sig.
1.	{N}	{SR}{SP}{RP} {S}{R}{P}	{D}	9.065	5	.107
2.	{N}	{SR}{RP}{S}{R} {P}	{D}	10.313	9	.326
3.	{N}	{SP}{RP}{S}{R} {P}	{D}	9.701	9	.375
4.	{N}	{SR}{SP}{S}{R} {P}	{D}	18.568	13	.137
5.	{N}	{RP}{S}{R}{P}	{D}	11.406	13	.577
6.	{N}	{SR}{S}{R}{P}	{D}	22.115	17	.180
7.	{N}	{SP}{S}{R}{P}	{D}	21.256	17	.215
8.	{N}	{S}{R}{P}	{D}	24.264	21	.280

acceptable fit to the data; however, the model also comes close to suggesting differences between observed and expected frequencies ($p = .107$). Models 2, 3, and 4 each contain two two-way interactions, with Model 4 fitting the data and offering an increase in degrees of freedom over Model 2, which also offers an acceptable fit. In comparing Model 3 with Models 2 and 4, it appears the two-way interaction between race and personal optimism could be absent in the data. Model 5, the first of three models containing just one interaction, supports that possibility, as it does not offer an acceptable fit, at least according to the Knoke and Burke (1980) criteria. Models 6 and 7 do fit the data and increase available degrees of freedom from 13 to 17. But Model 8 also offers an acceptable fit, indicating independence among the explanatory measures and increasing degrees of freedom to 21. In this case, one could predict national optimism by knowing the sex, race, and level of personal optimism among respondents, as well as an average amount of political discussion per week. No interactions proved necessary.

Tables 5.2a and 5.2b contain frequency data for males and females, respectively. The tables are similar to those describing data in the general log-linear model, offering observed and expected frequencies as well as four indicators of differences between the two. Examining standardized residuals for males, none reached 2.0, which Lawal (2003) identified as a threshold for outliers. Studying Table 5.2b, however, it appears that Black females who were neither pessimistic nor optimistic about their personal futures indicated slightly more national optimism than expected. As a rule, researchers should examine frequency data in a selected model before moving to an interpretation of parameter estimates.

Table 5.2a Descriptive statistics for logit log-linear models including sex (males), race, personal optimism, and frequency of political discussion as explanatory measures of national optimism

			Males						
Race	Personal Future	National Future	Observed Count (%)		Expected Count (%)		Residual	Standard Residual	Adjusted Residual	Deviance
White	Optimistic	Optimistic	128	68.4	128.745	68.8	-.745	-.118	-.208	-1.219
		Pessimistic	22	11.8	22.128	11.8	-.128	-.029	-.048	-.508
		Neither	37	19.8	36.128	19.3	.874	.162	.285	1.330
	Pessimistic	Optimistic	5	15.2	5.424	16.4	-.424	-.199	-.260	-.903
		Pessimistic	21	63.6	21.736	65.9	-.736	-.270	-.364	-1.202
		Neither	7	21.2	5.840	17.7	1.160	.529	.676	1.503
	Neither	Optimistic	23	31.9	23.586	32.8	-.586	-.147	-.213	-1.076
		Pessimistic	15	20.8	12.062	16.8	2.938	.927	1.281	2.557
		Neither	34	47.2	36.352	50.5	-2.352	-.554	-.798	-2.133
Black	Optimistic	Optimistic	43	76.8	42.635	76.1	.365	.114	.151	.856
		Pessimistic	2	3.6	3.737	6.7	-1.737	-.930	-1.218	-1.581
		Neither	11	19.6	9.628	17.2	1.372	.486	.643	1.712
	Pessimistic	Optimistic	0	0.0	1.011	12.6	-1.011	-1.076	-1.287	.000
		Pessimistic	5	62.5	4.636	58.0	.364	.260	.305	.869
		Neither	3	37.5	2.352	29.4	.648	.503	.576	1.208
	Neither	Optimistic	11	36.7	7.491	25.0	3.509	1.480	1.803	2.907
		Pessimistic	2	6.7	2.378	7.9	-.378	-.255	-.295	-.832
		Neither	17	56.7	20.131	67.1	-3.131	-1.216	-1.525	-2.397
Other Race	Optimistic	Optimistic	31	68.9	32.879	73.1	-1.879	-.631	-.854	-1.910
		Pessimistic	3	6.7	3.086	6.9	-.086	-.051	-.063	-.413
		Neither	11	24.4	9.035	20.1	1.965	.731	.961	2.081
	Pessimistic	Optimistic	2	50.0	.712	17.8	1.288	1.684	1.809	2.033
		Pessimistic	2	50.0	2.372	59.3	-.372	-.379	-.430	-.826
		Neither	0	0.0	.916	22.9	-.916	-1.090	-1.202	.000
	Neither	Optimistic	4	19.0	4.516	21.5	-.516	-.274	-.331	-.985
		Pessimistic	2	9.5	1.864	8.9	.136	.104	.129	.530
		Neither	15	71.4	14.619	69.9	.381	.181	.226	.878

Table 5.2b Descriptive statistics for logit log-linear models including sex (females), race, personal optimism, and frequency of political discussion as explanatory measures of national optimism

			Females								
Race	Personal Future	National Future	Observed Count	(%)	Expected Count	(%)	Residual	Standard Residual	Adjusted Residual	Deviance
White	Optimistic	Optimistic	177	70.8	178.294	71.3	-1.294	-.181	-.340	-1.606
		Pessimistic	31	12.4	29.966	12.0	1.034	.201	.361	1.450
		Neither	42	16.8	41.740	16.7	.260	.044	.074	.723
	Pessimistic	Optimistic	8	18.6	7.157	16.6	.843	.345	.494	1.335
		Pessimistic	28	65.1	27.938	65.0	.062	.020	.029	.352
		Neither	7	16.3	7.095	18.4	-.905	-.356	-.495	-1.305
	Neither	Optimistic	31	35.2	28.793	32.7	2.207	.501	.843	2.140
		Pessimistic	14	15.9	17.170	19.5	-3.170	-.853	-1.351	-2.391
		Neither	43	48.9	42.037	47.8	.963	.206	.382	1.396
Black	Optimistic	Optimistic	77	81.9	72.545	77.2	4.455	1.095	1.757	3.030
		Pessimistic	6	6.4	6.267	6.7	-.267	-.110	-.263	-.722
		Neither	11	11.7	15.188	16.2	-4.188	-1.174	-1.694	-2.664
	Pessimistic	Optimistic	0	0.0	.795	9.9	-.795	-.940	-1.104	.000
		Pessimistic	5	62.5	3.612	45.2	1.388	.986	1.185	1.803
		Neither	3	37.5	3.592	44.9	-.592	-.421	-.527	-1.040
	Neither	Optimistic	7	16.7	13.522	32.2	-6.522	-2.154	-3.044	-3.036
		Pessimistic	4	9.5	3.370	8.0	.630	.358	.426	1.171
		Neither	31	73.8	25.109	59.8	5.891	1.854	2.619	3.615
Other Race	Optimistic	Optimistic	29	70.7	29.902	72.9	-.902	-.317	-.410	-1.333
		Pessimistic	4	9.8	2.816	6.9	1.184	.731	.928	1.676
		Neither	8	19.5	8.282	20.2	-.282	-.110	-.138	-.745
	Pessimistic	Optimistic	1	50.0	.900	45.0	.100	.142	.175	.459
		Pessimistic	0	0.0	.706	35.3	-.706	-1.044	-1.178	.000
		Neither	1	50.0	.394	19.7	.606	1.076	1.140	1.364
	Neither	Optimistic	6	40.0	4.092	27.3	1.908	1.106	1.270	2.143
		Pessimistic	1	6.7	1.156	7.7	-.156	-.151	-.173	-.538
		Neither	8	53.3	9.753	65.0	-1.753	-.949	-1.124	-1.780

Logit analyses produce values for constants, as indicated in Table 5.3. Constants are not parameters and therefore do not contain standard errors, Z-values, significance levels, or confidence intervals. As a matter of practice, however, the values should be reported as part of a logit log-linear analysis. While published logit tables generally do not contain reference categories, Table 5.3 includes them for purposes of discussion.

Table 5.4 contains parameter estimates for the independence model reported in Table 5.1. Each explanatory measure contains at least one estimate as well as a reference category. To calculate odds ratios, one can exponentiate parameter estimates and then interpret the resulting figures in relation to 1.0, which indicates independence. As an example, the estimate for males and optimism, –.007, exponentiates to .993, indicating similar response patterns for males and females.

At the level of national pessimism, differences did not emerge between males and females, but in analyses of race, White respondents did differ from members of other races (the reference category for the independent variable). The table shows a parameter value of .757, which exponentiates to 2.13. Thus, for a White respondent, the odds of expressing a pessimistic attitude to a neither optimistic nor pessimistic attitude were approximately twice those same odds for an individual from another race. Returning to Tables 5.2a and 5.2b,

Table 5.3 Constant estimates for logit log-linear model including sex, race, personal optimism, and frequency of political discussion as explanatory measures of national optimism

Constant	*Estimate*
Male, White, Personally Optimistic	2.654
Male, White, Personally Pessimistic	1.382
Male, White, Neither Optimistic nor Pessimistic	3.090
Male, Black, Personally Optimistic	1.564
Male, Black, Personally Pessimistic	.520
Male, Black, Neither Optimistic nor Pessimistic	2.154
Male, Other Race, Personally Optimistic	1.439
Male, Other Race, Personally Pessimistic	–.088
Male, Other Race, Neither Optimistic nor Pessimistic	1.967
Female, White, Personally Optimistic	2.957
Female, White, Personally Pessimistic	1.636
Female, White, Neither Optimistic nor Pessimistic	3.442
Female, Black, Personally Optimistic	2.050
Female, Black, Personally Pessimistic	.273
Female, Black, Neither Optimistic nor Pessimistic	2.499
Female, Other Race, Personally Optimistic	1.360
Female, Other Race, Personally Pessimistic	–1.266
Female, Other Race, Neither Optimistic nor Pessimistic	1.523

collapsing the data across personal attitudes as well as sex indicates that while 19.5% of White respondents expressed pessimism, just 9.4% of those from other races indicated the same. Among Black respondents, 10.1% expressed pessimism, slightly higher than individuals from other races. The parameter estimate for Black respondents, .083, exponentiates to 1.09, indicating that the odds of a Black respondent expressing a pessimistic attitude to a neither pessimistic nor optimistic attitude were approximately 1.09 times the same odds for members of other races. With 1.0 indicating independence, the two groups did not differ significantly.

Personal optimism also explained attitudes at the national level. For example, for an individual who expressed personal optimism, the odds of expressing national optimism to neither optimism nor pessimism were $\exp(2.133) = 8.44$ times the same odds for an individual who indicated neither (personal) optimism nor pessimism. For an individual who expressed personal pessimism, the odds of expressing national pessimism to neither optimism or pessimism were $\exp(2.296) = 9.93$ times the same odds for an individual who indicated neither (personal) optimism nor pessimism. Interestingly, individuals expressing personal optimism also appeared more likely to indicate national pessimism.

Parameter estimates for the relationship between the continuous covariate, frequency of political discussion, and the three levels of national optimism, appear at the bottom of Table 5.4. In this case, the covariate, included for didactic purposes, did not show significance as a control measure. Notably, one of the primary differences between the logit log-linear model and logistic regression analysis involves the treatment of continuous measures; in log-linear analyses, such measures must be treated as covariates, but as subsequent chapters in the current text demonstrate, logistic regression models accommodate interval-level explanatory measures. The logit log-linear model accommodates more than one response measure, though, as discussed in the following example.

Logit Model with Two Response Measures

Drawing on data gathered in the 2012 Monitoring the Future study (Johnston et al. 2012), this section of the chapter reviews a logit log-linear model containing two dependent variables. The analysis examines the explanatory effects of three categorical factors – sex (S), race (R), and whether respondents had ever been suspended from school (P) – in addition to a covariate measuring frequency of exposure to televised antidrug advertisements (V), on disapproval of alcohol consumption (A), and disapproval of marijuana use (M). White, Black, and Hispanic respondents indicated whether (1) they did not disapprove, (2) they disapproved, or (3) they strongly disapproved of individuals

Table 5.4 Parameter estimates for logit log-linear model including sex, race, personal optimism, and frequency of political discussion as explanatory measures of national optimism

Parameter	Estimate	SE	Z	Sig.	95% Confidence Interval	
					Lower Bound	Upper Bound
National Optimism	−.687	.503	−1.365	.172	−1.674	.299
National Pessimism	−1.379	.580	−2.375	.018	−2.516	−.241
National Neither						
National Optimism × Male	−.007	.170	−.039	.969	−.340	.327
National Optimism × Female						
National Pessimism × Male	−.006	.241	−.027	.979	−.479	.466
National Pessimism × Female						
National Neither × Male						
National Neither × Female						
National Optimism × White	.030	.265	.115	.908	−.489	.550
National Optimism × Black	−.100	.332	−.301	.764	−.752	.552
National Optimism × Other Race						
National Pessimism × White	.757	.372	2.039	.041	.029	1.486
National Pessimism × Black	.083	.434	.192	.848	−.768	.934
National Pessimism × Other Race						

(Continued)

Table 5.4 (Continued)

Parameter	Estimate	SE	Z	Sig.	95% Confidence Interval	
					Lower Bound	Upper Bound
National Neither × White						
National Neither × Black						
National Neither × Other Race						
National Optimism × Personal Optimism	2.133	.235	9.056	.000	1.671	2.594
National Optimism × Personal Pessimism	.285	.369	.773	.440	−.438	1.009
National Optimism × Personal Neither						
National Pessimism × Personal Optimism	1.043	.294	3.546	.000	.466	1.619
National Pessimism × Personal Pessimism	2.296	.319	7.205	.000	1.671	2.920
National Pessimism × Personal Neither						
National Neither × Personal Optimism						
National Neither × Personal Pessimism						
National Neither × Personal Neither						
National Optimism × Political Discussion	.312	.209	1.498	.134	−.096	.721
National Pessimism × Political Discussion	.009	.138	.066	.948	−.261	.279
National Neither × Political Discussion	.335	.183	1.828	.068	−.024	.695

over 18 consuming one to two drinks of alcohol per day and smoking marijuana on a regular basis. Of the models contained in Table 5.5, just two fit the data based on the Knoke and Burke (1980) criteria. Model 3 contains interactions between sex and race as well as sex and school suspension, and Model 8 indicates main effects only. The difference between the two models, 28.247 with 24 degrees of freedom, is not significant, and therefore the simpler model is preferred. Table 5.6 contains constants for Model 8, and Tables 5.7a and 5.7b contain parameter estimates.

Table 5.5   Logit log-linear models including sex, race, school suspension, and exposure to antidrug advertising as explanatory measures of alcohol and marijuana disapproval

	Dependent Variables	*Independent Variables*	*Covariate*	G^2	*df*	*Sig.*
1.	{A}{M}	{SR}{SP}{RP}{S}{R}{P}	{V}	7.367	7	.498
2.	{A}{M}	{SP}{RP}{S}{R}{P}	{V}	17.731	23	.772
3.	{A}{M}	{SR}{SP}{S}{R}{P}	{V}	28.656	23	.192
4.	{A}{M}	{SR}{RP}{S}{R}{P}	{V}	15.588	15	.410
5.	{A}{M}	{SR}{S}{R}{P}	{V}	32.078	31	.413
6.	{A}{M}	{RP}{S}{R}{P}	{V}	25.748	31	.733
7.	{A}{M}	{SP}{S}{R}{P}	{V}	53.298	39	.063
8.	{A}{M}	{S}{R}{P}	{V}	56.903	47	.153

Table 5.6   Constant estimates for logit log-linear model containing sex, race, school suspension, and exposure to antidrug advertising as explanatory measures of alcohol and marijuana disapproval

Constant	*Estimate*
Male × White × No Suspension	1.115
Male × White × Suspension	0.333
Male × Black × No Suspension	3.075
Male × Black × Suspension	1.361
Male × Hispanic × No Suspension	1.470
Male × Hispanic × Suspension	0.364
Female × White × No Suspension	1.709
Female × White × Suspension	0.782
Female × Black × No Suspension	3.516
Female × Black × Suspension	0.843
Female × Hispanic × No Suspension	2.097
Female × Hispanic × Suspension	0.032

Table 5.7a Parameter estimates for logit log-linear model including sex, race, school suspension, and exposure to antidrug advertising as explanatory measures of alcohol and marijuana disapproval

Parameter	Estimate	SE	Z	Sig.	95% Confidence Interval	
					Lower Bound	Upper Bound
Alcohol Do Not Disapprove	−.740	1.096	−.675	.500	−2.888	1.409
Alcohol Disapprove	1.043	1.499	.696	.486	−1.894	3.981
Marijuana Do Not Disapprove	−.707	1.163	−.608	.543	−2.986	1.571
Marijuana Disapprove	−1.677	1.587	−1.057	.291	−4.787	1.433
Alcohol Do Not Disapprove × Marijuana Do Not Disapprove	1.693	1.615	1.048	.294	−1.472	4.857
Alcohol Do Not Disapprove × Marijuana Disapprove	.918	2.045	.449	.653	−3.089	4.926
Alcohol Disapprove × Marijuana Do Not Disapprove	1.052	1.895	.555	.579	−2.662	4.766
Alcohol Disapprove × Marijuana Disapprove	3.389	2.352	1.436	.151	−1.231	7.988
Alcohol Do Not Disapprove × Male	.215	.234	.918	.359	−.244	.673
Alcohol Disapprove × Male	.082	.237	.345	.730	−.382	.546
Marijuana Do Not Disapprove × Male	.703	.343	2.052	.040	.032	1.375
Marijuana Disapprove × Male	.086	.243	.355	.723	−.390	.563
Alcohol Do Not Disapprove ×	.272	.423	.642	.521	−.558	1.102

Marijuana Do Not Disapprove × Male Alcohol Do Not Disapprove × Marijuana Disapprove × Male	.272	.367	.741	.459	−.447	.990
Alcohol Disapprove × Marijuana Do Not Disapprove × Male	−.269	.431	−.624	.533	−1.115	.576
Alcohol Disapprove × Marijuana Disapprove × Male	−.093	.340	−.274	.784	−.760	.574
Alcohol Do Not Disapprove × White	.220	.553	.397	.691	−.864	1.303
Alcohol Do Not Disapprove × Black	1.247	.506	2.463	.014	.255	2.239
Alcohol Disapprove × White	−.630	.346	−1.821	.069	−1.308	.048
Alcohol Disapprove × Black	.118	.319	.371	.711	−.507	.744
Marijuana Do Not Disapprove × White	−.092	.488	−.189	.850	−1.048	.864
Marijuana Do Not Disapprove × Black	−.502	.374	−1.343	.179	−1.234	.230
Marijuana Disapprove × White	−1.247	.509	−2.447	.014	−2.245	−.248
Marijuana Disapprove × Black	−.137	.323	−.423	.673	−.770	.497
Alcohol Do Not Disapprove × Marijuana Do Not Disapprove × White	−.601	.782	−.768	.442	−2.135	.933
Alcohol Do Not Disapprove × Marijuana Do Not Disapprove × Black	−.299	.644	−.465	.642	−1.561	.963

Table 5.7b Parameter estimates for logit log-linear model including sex, race, school suspension, and exposure to antidrug advertising as explanatory measures of alcohol and marijuana disapproval

Parameter	Estimate	SE	Z	Sig.	95% Confidence Interval	
					Lower Bound	Upper Bound
Alcohol Do Not Disapprove × Marijuana Disapprove × White	.579	.880	.658	.510	−1.146	2.304
Alcohol Do Not Disapprove × Marijuana Disapprove × Black	−.872	.651	−1.340	.180	−2.147	.403
Alcohol Disapprove × Marijuana Do Not Disapprove × White	.541	.623	.869	.385	−.679	1.762
Alcohol Disapprove × Marijuana Do Not Disapprove × Black	.605	.528	1.146	.252	−.429	1.639
Alcohol Disapprove × Marijuana Disapprove × White	1.791	.622	2.880	.004	.572	3.010
Alcohol Disapprove × Marijuana Disapprove × Black	.093	.463	.200	.841	−.815	1.001
Alcohol Do Not Disapprove × No Suspend	−.506	.340	−1.491	.136	−1.172	.159
Alcohol Disapprove × No Suspend	−.001	.295	−.003	.998	−.579	.577
Marijuana Do Not Disapprove × No Suspend	−1.241	.344	−3.605	.000	−1.915	−.566
Marijuana Disapprove × No Suspend	−.420	.313	−1.343	.179	−1.033	.193
Alcohol Do Not Disapprove × Marijuana	.770	.481	1.601	.109	−.172	1.712

Do Not Disapprove × No Suspend Alcohol Do Not Disapprove × Marijuana Disapprove × No Suspend	.308	.476	.646	.518	−.626	1.241
Alcohol Disapprove × Marijuana Do Not Disapprove × No Suspend	.017	.460	.037	.971	−.886	.919
Alcohol Disapprove × Marijuana Disapprove × No Suspend	−.111	.423	−.262	.793	−.939	.718
Alcohol Do Not Disapprove × Antidrug	.111	.229	.486	.627	−.338	.560
Alcohol Disapprove × Antidrug	.001	.489	.002	.998	−.957	.959
Alcohol Strongly Disapprove × Antidrug	.562	.248	2.268	.023	.076	1.048
Marijuana Do Not Disapprove × Antidrug	−.113	.387	−.293	.770	−.871	.645
Marijuana Disapprove × Antidrug	.360	.538	.669	.503	−.695	1.415
Alcohol Do Not Disapprove × Marijuana Do Not Disapprove × Antidrug	.278	.533	.521	.603	−.767	1.323
Alcohol Do Not Disapprove × Marijuana Disapprove × Antidrug	.341	.714	.478	.633	−1.058	1.740
Alcohol Disapprove × Marijuana Do Not Disapprove × Antidrug	.092	.644	.143	.886	−1.170	1.355
Alcohol Disapprove × Marijuana Disapprove × Antidrug	−.782	.838	−.934	.350	−2.424	.860

As evidenced in these tables, the introduction of a second dependent variable increases the number of parameter estimates substantially. In Table 5.7a (which, like Table 5.7b, does not include reference categories), two sets of associations between the two dependent variables appear. The first set shows individuals who did not disapprove of alcohol consumption at each level of the marijuana measure, and the second set shows individuals who disapproved of alcohol use at each level of the marijuana variable. In this analysis, for an individual who did not disapprove of alcohol use, the odds of not disapproving of marijuana use were $\exp(1.693) = 5.44$ times the same odds for an individual who strongly disapproved of alcohol use; however, as indicated in the table, this association was not statistically significant. Additional odds ratios concerning the dependent variables can be calculated by exponentiating parameter estimates in the two sets of associations.

Regarding sex and marijuana use, the odds of a male indicating no disapproval to strong disapproval were $\exp(.703) = 2.02$ times the same odds for a female, and the relationship showed significance at $p < .05$. Concerning race and attitudes toward marijuana use, the odds of a White respondent indicating disapproval to strong disapproval were $\exp(-1.247) = .29$ times the same odds for an Hispanic respondent. In analyses addressing alcohol use, the odds of a Black respondent not disapproving to strongly disapproving were $\exp(1.247) = 3.48$ times the same odds for Hispanics.

Looking to Table 5.7b, one observes a significant three-way interaction among White respondents, those who disapproved of alcohol use, and those who disapproved of marijuana use. This interaction indicates that for a White respondent, the odds of disapproving to strongly disapproving of alcohol use and the odds of disapproving to strongly disapproving of marijuana use, were $\exp(1.791) = 6.0$ times the same odds for Hispanic respondents. Toward the bottom of Table 5.7b, one observes parameter estimates for the two dependent measures and the covariate measuring exposure to antidrug campaigns. The odds of an individual exposed to antidrug communications strongly disapproving of alcohol use were $\exp(.562) = 1.75$ times the same odds for an individual strongly disapproving of marijuana use (the reference category).

SPSS Example

This section of the chapter offers instructions for conducting a logit log-linear analysis in SPSS (Norusis 2005: 25–42). In doing so, the chapter draws on data gathered in the 2011 National Survey on Drug Use and Health ($N = 17,721$), with four categorical factors predicting attitudes toward peer alcohol consumption. Explanatory measures included dichotomous indictors of (1) sex, (2)

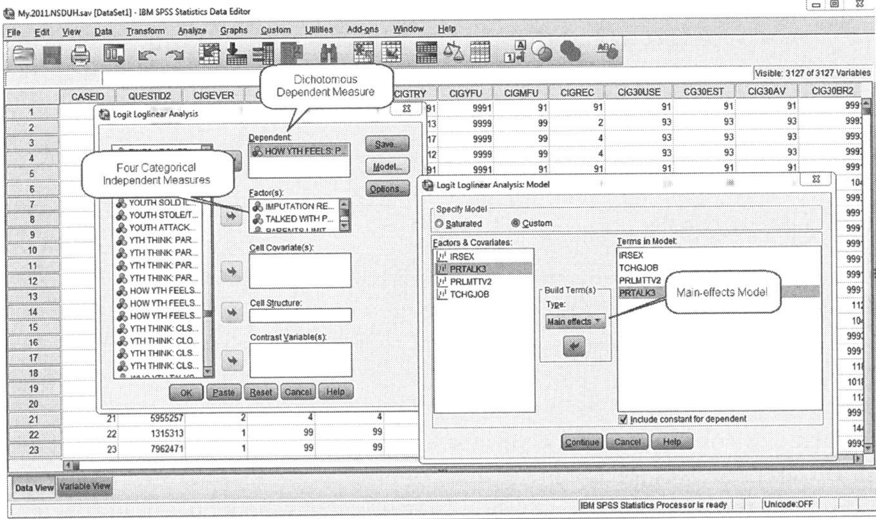

Figure 5.1   Screenshot of SPSS Logit Loglinear Analysis. Source: SPSS® Reprints Courtesy of International Business Machines Corporation, © 2014 International Business Machines Corporation

whether students had discussed the dangers of drugs and alcohol with their parents, (3) whether parents had limited the amount of time in which students could watch television, and (4) whether students had received positive feedback from teachers. NSDUH researchers had created a dichotomous variable asking respondents about peers consuming 1–2 drinks each day, with Disapprove and Strongly Disapprove collapsed to form one level and Neither Approve nor Disapprove collapsed to form a second category. For this analysis, then, all variables are dichotomous.

Figure 5.1 contains a screenshot of logit log-linear options in SPSS. To begin a logit analysis, a researcher should click on the Analysis menu, select Loglinear, and then Logit. Doing so opens a window in which the researcher can choose independent and dependent variables as well as covariates. As with general log-linear analysis in SPSS, the Model window allows one to fit both saturated and custom models to the data; the window in Figure 5.1 shows a main-effects model based on independent and dependent measures selected at the previous step. As it happens, this model offered an acceptable fit to the data, with a log-likelihood value of 12.328 and 11 degrees of freedom yielding a p-value of .340. The Options window allows a researcher to select parameter estimates as well as any plots that might inform a study.

Table 5.8   SPSS goodness-of-fit display for logit log-linear model containing sex, parental communication, limited television viewing, and teacher encouragement as predictors of attitudes toward peer alcohol consumption

Goodness-of-Fit Tests [a,b]

Value	df	Sig.	
Likelihood Ratio	12.328	11	.340
Pearson Chi-Square	12.201	11	.349

[a] Model: Multinomial Logit
[b] Design: Constant + YFLADLY2 + YFLADLY2 * IRSEX + YFLADLY2 * TCHGJOB + YFLADLY2 * PRLMTTV2 + YFLADLY2 * PRTALK3

Table 5.8 contains goodness-of-fit statistics from the SPSS output. The table shows similar values for G^2 and χ^2 and confirms a multinomial logit model has been fit to the data. The Design shows a main-effects model has been tested.

Table 5.9 contains cell counts and residuals for the main-effects model, as presented in SPSS. As Table 5.9 illustrates, categories for explanatory measures begin at the left margin and move across to the dichotomous response variable. Observed and expected values appear relatively close, and no standardized residuals exceed 2.0. Table 5.10 contains constants for the model.

Table 5.11 contains parameter estimates for the main-effects model. Relative to the tables shown earlier in the chapter, this table does not display as many estimates because the explanatory and response measures were all dichotomous. Concerning the latter, respondents were $\exp(1.313) = 3.72$ times as likely to indicate disapproval of alcohol use as they were to indicate neither approval nor disapproval. Regarding the explanatory factors, for a male respondent, the odds of indicating disapproval to neither approval nor disapproval were $\exp(-.318) = .73$ times the same odds for a female. For a respondent who received scholastic encouragement from a teacher, the odds of indicating disapproval to neither approval nor disapproval were $\exp(.553) = 1.74$ times the same odds for a respondent who did not receive teacher encouragement. For a respondent whose parents limited television viewing, the odds of disapproving to neither approving nor disapproving were $\exp(.638) = 1.89$ times the same odds for respondents whose parents did not limit television viewing. Lastly, for an individual who had discussed the dangers of drugs and alcohol with parents, the odds of disapproving to neither approving nor disapproving were $\exp(.473) = 1.60$ times the same odds for an individual who had not discussed the dangers with parents. As indicated in Table 5.11, each explanatory factor showed statistical significance, and the odds ratios just reviewed indicate the strength of variable associations.

Table 5.9 Cell Counts and Residuals for logit log-linear model containing sex, parental communication, limited television viewing, and teacher encouragement as predictors of attitudes toward peer alcohol consumption

Teacher Encourages	Parents Limit TV	Parents Comm	Sex	Peers Drink	Observed Count	(%)	Expected Count	(%)	Residual	Standard Residual	Adjusted Residual	Deviance
Yes	Yes	Yes	Male	Disapprove	1832	92.9	1842.868	93.5	-10.868	-.989	-1.244	-4.655
				No Disapproval	140	7.1	129.132	6.5	10.868	.989	1.244	4.757
			Female	Disapprove	1894	96.0	1877.324	95.2	16.676	1.748	2.113	5.788
				No Disapproval	79	4.0	95.676	4.8	-16.676	-1.748	-2.113	-5.501
		No	Male	Disapprove	898	90.0	895.307	89.9	.693	.073	.088	1.178
				No Disapproval	100	10.0	100.693	10.1	-.693	-.073	-.088	-1.175
			Female	Disapprove	787	91.6	794.047	92.4	-7.047	-.910	-1.038	-3.746
				No Disapproval	72	8.4	64.953	7.6	7.047	.910	1.038	3.851
	No	Yes	Male	Disapprove	2136	87.7	2150.000	88.3	-14.000	-.883	-1.268	-5.283
				No Disapproval	299	12.3	285.000	11.7	14.000	.883	1.268	5.355
			Female	Disapprove	2106	91.6	2095.928	91.2	10.072	.742	.993	4.493
				No Disapproval	192	8.4	202.072	8.8	-10.072	-.742	-.993	-4.431
		No	Male	Disapprove	1560	83.6	1538.638	82.5	21.362	1.300	1.904	6.559
				No Disapproval	306	16.4	327.362	17.5	-21.362	-1.300	-1.904	-6.426
			Female	Disapprove	1412	85.6	1428.888	86.6	-16.888	-1.220	-1.640	-5.794
				No Disapproval	238	14.4	221.112	13.4	16.888	1.220	1.640	5.919

(*Continued*)

Table 5.9 (Continued)

Teacher Encourages	Parents Limit TV	Parents Comm	Sex	Peers Drink	Observed Count	(%)	Expected Count	(%)	Residual	Standard Residual	Adjusted Residual	Deviance
No	Yes	Yes	Male	Disapprove	286	89.7	284.356	89.1	1.644	.296	.321	1.816
				No Disapproval	33	10.3	34.644	10.9	−1.644	−.296	−.321	−1.791
			Female	Disapprove	249	91.9	248.941	91.9	.059	.013	.014	.343
				No Disapproval	22	8.1	22.059	8.1	−.059	−.013	−.014	−.343
		No	Male	Disapprove	182	83.1	183.180	83.6	−1.180	−.216	−.233	−1.534
				No Disapproval	37	16.9	35.820	16.4	1.180	.216	.233	1.549
			Female	Disapprove	183	87.6	182.977	87.5	.023	.005	.005	.217
				No Disapproval	26	12.4	26.023	12.5	−.023	−.005	−.005	−.217
	No	Yes	Male	Disapprove	469	79.8	477.864	81.3	−8.864	−.937	−1.119	−4.191
				No Disapproval	119	20.2	110.136	18.7	8.864	.937	1.119	4.292
			Female	Disapprove	520	86.5	514.718	85.6	5.282	.614	.714	3.258
				No Disapproval	81	13.5	86.282	14.4	−5.282	−.614	−.714	−3.199
		No	Male	Disapprove	520	74.6	508.787	73.0	11.213	.957	1.221	4.761
				No Disapproval	177	25.4	188.213	27.0	−11.213	−.957	−1.221	−4.663
			Female	Disapprove	597	77.7	605.176	78.8	−8.176	−.722	−.918	−4.030
				No Disapproval	171	22.3	162.824	21.2	8.176	.722	.918	4.094

Table 5.10 Constant estimates for logit log-linear model containing sex, parental communication, limited television viewing, and teacher encouragement as predictors of attitudes toward peer alcohol consumption

Constant	Estimate
Teacher Encouragement × TV Limit × Parental Communication × Male	4.861
Teacher Encouragement × TV Limit × Parental Communication × Female	4.561
Teacher Encouragement × TV Limit × No Parental Communication × Male	4.612
Teacher Encouragement × TV Limit × No Parental Communication × Female	4.174
Teacher Encouragement × No TV Limit × Parental Communication × Male	5.652
Teacher Encouragement × No TV Limit × Parental Communication × Female	5.309
Teacher Encouragement × No TV Limit × No Parental Communication × Male	5.791
Teacher Encouragement × No TV Limit × No Parental Communication × Female	5.399
No Teacher Encouragement × TV Limit × Parental Communication × Male	3.545
No Teacher Encouragement × TV Limit × Parental Communication × Female	3.094
No Teacher Encouragement × TV Limit × No Parental Communication × Male	3.579
No Teacher Encouragement × TV Limit × No Parental Communication × Female	3.259
No Teacher Encouragement × No TV Limit × Parental Communication × Male	4.702
No Teacher Encouragement × No TV Limit × Parental Communication × Female	4.458
No Teacher Encouragement × No TV Limit × No Parental Communication × Male	5.238
No Teacher Encouragement × No TV Limit × No Parental Communication × Female	5.093

Table 5.11 Parameter estimates for logit log-linear model containing sex, parental communication, limited television viewing, and teacher encouragement as predictors of attitudes toward peer alcohol consumption

Parameter	Estimate	SE	Z	Sig.	95% CI Lower	95% CI Upper
Disapprove of Alcohol Use	1.313	.055	24.056	.000	1.206	1.420
Do Not Disapprove of Alcohol Use						
Disapprove of Alcohol Use × Male	−.318	.048	−6.670	.000	−.412	−.225
Disapprove of Alcohol Use × Female						
Do Not Disapprove of Alcohol Use × Male						
Do Not Disapprove of Alcohol Use × Female						
Disapprove of Alcohol Use × Teacher Encourage	.553	.052	10.601	.000	.451	.655
Disapprove of Alcohol use × No Teacher Encourage						
Do Not Disapprove of Alcohol Use × Teacher Encourage						
Do Not Disapprove of Alcohol Use × No Teacher Encourage						
Disapprove of Alcohol Use × TV Limit	.638	.054	11.737	.000	.531	.744
Disapprove of Alcohol Use × No TV Limit						
Do Not Disapprove of Alcohol Use × TV limit						
Do Not Disapprove of Alcohol Use × No TV Limit						
Disapprove of Alcohol Use × Parental Communication	.473	.048	9.917	.000	.380	.567
Disapprove of Alcohol Use × No Parental Communication						
Do Not Disapprove of Alcohol Use × Parental Communication						
Do Not Disapprove of Alcohol Use × No Parental Communication						

Correspondence Analysis

In recent years, communication scholars have used correspondence analysis to graph and display categorical data (see Brito 2012, Hovden 2014, Sonnet 2010). Developed and widely used in France (Benzécri 1973), correspondence analysis serves as an effective complement to general and logit log-linear modeling (see Van der Heijden and de Leeuw 1985, Van der Heijden, de Falguerolles, and de Leeuw 1989), transforming values in a contingency table into coordinates for mapping in conceptual space (Greenacre 1984). Basic correspondence analysis includes an X and a Y measure, and multiple correspondence analysis includes more than two variables. Categories with similar distributions tend to cluster in conceptual space, whereas categories with dissimilar patterns appear comparatively dispersed (Clausen 1998). In this regard, correspondence analysis resembles computer-assisted concept-mapping procedures (see, e.g., Miller, Andsager, and Riechert 1998).

Statisticians regard correspondence analysis as a special form of canonical correlation, with coordinates considered analogous to values derived in a principal components analysis (Everitt 1992, 48–53). In fact, statisticians often discuss correspondence analysis in the context of multidimensional scaling, which also resembles factor analysis and describes variable structures (Nunnally and Bernstein 1994, 621–622). In SPSS, correspondence analysis requires an add-on module, but to provide readers with a similar representation, Figure 5.2 contains an SPSS map of 11 ordinal and discrete interval measures extracted from the 2011 National Survey on Drug Use and Health (United States Department of Health and Human Services 2011). This map, produced in the SPSS multidimensional scaling program, includes variables addressing the extent to which adolescents had argued with their parents, argued with or fought with students at school, received positive feedback from a teacher, received positive feedback from a parent, and received homework assistance from a parent. Variables also indicated the frequency with which adolescents had been asked to complete chores, how often schoolwork appeared meaningful, how students felt about attending school, and how many school-, community-, and faith-based activities adolescents had participated in during the previous year.

As shown in Figure 5.2, the lower left section contains the three variables measuring activity participation, with faith-based activities not as closely related as school and community activities. The lower right section indicates that arguing with parents did not associate with any measure, while most of the other variables clustered toward the top of the figure. Although this map came from the SPSS multidimensional scaling program, it resembles the type of figure produced for correspondence analysis in the add-on module for categorical data. The map allows a researcher to visually inspect the data and consider the structure of variable relationships.

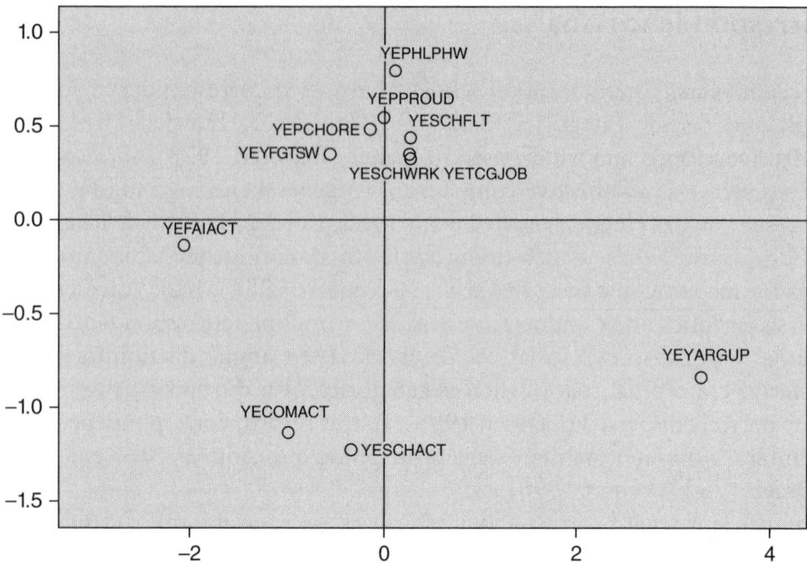

Figure 5.2   SPSS map of eleven ordinal variables. Source: SPSS® Reprints Courtesy of International Business Machines Corporation, © 2014 International Business Machines Corporation

Chapter Summary

This chapter has addressed the logit log-linear model, which functions as a categorical analog to ordinary least squares regression. The model allows researchers to examine the explanatory effects of categorical factors and continuous covariates on one or more categorical dependent variables. Like the general log-linear model, logit analysis produces parameter estimates in the form of log odds, which can be exponentiated to form odds ratios. Logit log-linear analyses can be conducted in SPSS, which offers both general and logit programs, and an SPSS add-on module facilitates correspondence analysis and data visualization.

Chapter Exercises

1.   Based on the material covered in the chapter as well as the horse-racing study (Denham 2014), use the Logit Loglinear Analysis program in SPSS to analyze the explanatory effects of time period (Before/After investigative report) and news outlet (*Los Angeles Times*, *The New York Times*, *Washington Post*, aggregated broadcast outlets) on whether news reports referenced an injured or deceased horse. Report first on whether the two-factor model fits based on

the Knoke and Burke (1980) criteria, and then report the parameter estimates and standard errors in addition to Z values, significance levels, and 95% confidence intervals. Exponentiate the parameter estimates and report the associated odds ratios, offering a brief summary of what the findings suggest about time period, news outlet, and references to injured or deceased horses. (See instructions for SPSS weighting procedure in Chapter 4 exercises.)

	News Outlet	*Reference to Injured/ Deceased Horse*	*Frequency*
Period 1	Los Angeles Times	Yes	13
		No	60
	The New York Times	Yes	38
		No	139
	Washington Post	Yes	20
		No	34
	Broadcasts	Yes	3
		No	35
Period 2	Los Angeles Times	Yes	32
		No	64
	The New York Times	Yes	52
		No	91
	Washington Post	Yes	14
		No	38
	Broadcasts	Yes	16
		No	50

2. Use the Logit Loglinear Analysis program in SPSS to analyze the explanatory effects of sex (Male, Female) and whether adolescents had attended a school lecture or film about substance abuse (Yes, No) on whether adolescents had discussed the dangers of alcohol and drug abuse with their parents (Yes, No). Report first on whether the two-factor model fits based on the Knoke and Burke (1980) criteria, and then report the parameter estimates and standard errors in addition to Z values, significance levels, and 95% confidence intervals. Exponentiate the parameter estimates and report the associated odds ratios, offering a brief summary of what the findings suggest about sex, school-based antidrug efforts, and communication with parents about substance abuse.

Sex	*Attend Film or Lecture*	*Talked with Parent(s) about Substance Use*	*Frequency*
Male	Yes	Yes	3,706
		No	2,088
	No	Yes	1,629
		No	1,720

Sex	Attend Film or Lecture	Talked with Parent(s) about Substance Use	Frequency
Female	Yes	Yes	3,716
		No	2,007
	No	Yes	1,438
		No	1,490

References

Agresti, Alan. 1990. *Categorical Data Analysis.* New York: John Wiley & Sons.

Aldrich, John H., and Forrest D. Nelson. 1984. *Linear Probability, Logit, and Probit Models.* Newbury Park, CA: Sage.

Allison, Paul D. 1999. "Comparing Logit and Probit Coefficients Across Groups." *Sociological Methods and Research*, 28: 186–208. DOI:10.1177/0049124199028 002003.

Benzécri, J. P. 1973. *L'analyse des Données.* Paris: Dunod.

Berkson, Joseph. 1944. "Application of the Logistic Function to Bio-Assay." *Journal of the American Statistical Association*, 39: 357–365.

Bishop, Yvonne M. M. 1969. "Full Contingency Tables, Logits, and Split Contingency Tables." *Biometrics*, 25: 119–128.

Borooah, Vani K. 2002. *Logit and Probit: Ordered and Multinomial Models.* Thousand Oaks, CA: Sage.

Brito, Pedro Quelhas. 2012. "Teen Conceptualization of Digital Technologies." *New Media & Society*, 14: 513–532. DOI:10.1177/1461444811420822.

Clausen, Sten-Erik. 1998. *Applied Correspondence Analysis: An Introduction.* Thousand Oaks, CA: Sage.

Coventry, Barbara Thomas. 2004. "On the Sidelines: Sex and Racial Segregation in Television Sports Broadcasting." *Sociology of Sport Journal*, 21: 322–341.

Davis, James A. 1978. "Hierarchical Models for Significance Tests in Multiway Contingency Tables: An Exegesis of Goodman's Recent Papers." In Leo A. Goodman, *Analyzing Qualitative/Categorical Data: Log-linear Models and Latent-Structure Analysis* (pp. 233–280). Cambridge, MA: Abt Books.

DeMaris, Alfred. 1992. *Logit Modeling: Practical Applications.* Newbury Park, CA: Sage.

Denham, Bryan E. 2009. "Determinants of Anabolic-Androgenic Steroid Risk Perceptions in Youth Populations: A Multivariate Analysis." *Journal of Health & Social Behavior*, 50: 277–292. DOI:10.1177/002214650905000303.

Denham, Bryan E. 2014. "Intermedia Attribute Agenda-Setting in the New York Times: The Case of Animal Abuse in U.S. Horse Racing." *Journalism & Mass Communication Quarterly*, 91: 17–37. DOI:10.1177/1077699013514415.

Everitt, B. S. 1992. *The Analysis of Contingency Tables*, 2nd ed. Boca Raton, FL: CRC Press.

Fienberg, Stephen E. 2007. *The Analysis of Cross-Classified Categorical Data.* New York: Springer.

Goodman, Leo A. 1970. "The Multivariate Analysis of Qualitative Data: Interactions among Multiple Classifications." *Journal of the American Statistical Association*, 65: 226–256.

Greenacre, M. J. 1984. *Theory and Applications of Correspondence Analysis.* New York: Academic Press.

Haberman, Shelby J. 1978. *Analysis of Qualitative Data.* New York: Academic Press.

Hagle, Timothy M., and Glenn E. Mitchell II. 1992. "Goodness-of-Fit Measures for Probit and Logit." *American Journal of Political Science*, 36: 762–784.

Hovden, Jan Fredrik. 2014. "To Intervene or be Neutral, to Investigate or Entertain?" *Journalism Practice*, 8: 646–659. DOI:10.1080/17512786.2014.894332.

Huang, Chi, and Todd G. Shields. 2000. "Interpretation and Interaction Effects in Logit and Probit Analyses." *American Politics Quarterly*, 28: 80–95. DOI:10.1177/1532673X00028001005.

Johnston, Lloyd D., Jerald G. Bachman, Patrick M. O'Malley, and John Schulenberg. 2012. Monitoring the Future: A Continuing Study of American Youth. Funded by National Institute on Drug Abuse. Institutional for Social Research, University of Michigan.

Knoke, David, and Peter J. Burke. 1980. *Log-linear Models.* Beverly Hills, CA: Sage.

Lawal, Bayo. 2003. *Categorical Data Analysis with SAS and SPSS Applications.* Mahwah, NJ: Erlbaum.

Liao, Tim Futing. 1994. *Interpreting Probability Models: Logit, Probit, and Other Generalized Linear Models.* Thousand Oaks, CA: Sage.

Magidson, Jay. 1978. "An Illustrative Comparison of Goodman's Approach to Logit Analysis with Dummy Variable Regression Analysis." In Leo A. Goodman, *Analyzing Qualitative/Categorical Data: Log-linear Models and Latent-Structure Analysis* (pp. 27–53). Cambridge, MA: Abt Books.

McFadden, Daniel. 1973. "Conditional Logit Analysis of Qualitative Choice Behavior." In P. Zarembka (Ed.), *Frontiers in Econometrics* (pp. 105–142). New York: Academic Press.

Miller, M. Mark, Julie L. Andsager, and Bonnie P. Riechert. 1998. "Framing the Candidates in Presidential Primaries: Issues and Images in Press Releases and News Coverage." *Journalism & Mass Communication Quarterly*, 75: 312–324. DOI:10.1177/107769909807500207.

Mulder, Ronald. 1981. "A Log-Linear Analysis of Media Credibility." *Journalism Quarterly*, 58: 635–638.

Norusis, Marija J. 2005. *SPSS 14.0 Advanced Statistical Procedures Companion.* Upper Saddle River, NJ: Prentice Hall.

Nunnally, Jum C., and Ira H. Bernstein. 1994. *Psychometric Theory*, 3rd ed. New York: McGraw-Hill.

Patterson, Miles L., Yuichi Iizuka, Mark E. Tubbs, Jennifer Ansel, Masao Tsutsumi, and Jackie Anson. 2007. "Passing Encounters East and West: Comparing Japanese and American Pedestrian Interactions." *Journal of Nonverbal Behavior*, 31: 155–166. DOI:10.1077/s10919-007-0028-4.

Patterson, Miles L., and Mark E. Tubbs. 2005. "Through a Glass Darkly: Effects of Smiling and Visibility on Recognition and Avoidance in Passing Encounters." *Western Journal of Communication*, 69: 219–231. DOI:10.1080/10570310500202389.

Powers, Daniel A., and Yu Xie. 2000. *Statistical Methods for Categorical Data Analysis*. San Diego, CA: Academic Press.

Siminoff, Laura A., Heather M. Traino, and Nahida H. Gordon. 2011. "An Exploratory Study of Relational, Persuasive, and Nonverbal Communication in Requests for Tissue Donation." *Journal of Health Communication*, 16: 955–975. DOI:10.1080/10810730.2011.561908.

Sonnet, John. 2010. "Climates of Risk: A Field Analysis of Global Climate Change in U.S. Media Discourse, 1997–2004." *Public Understanding of Science*, 19: 698–716. DOI:10.1177/0963662509346368.

The American National Election Studies. 2008. American National Election Study: ANES Pre- and Post-Election Survey. ICPSR25383-v2. Ann Arbor, MI: Inter-university Consortium for Political and Social Research [distributor], 2012-08-30. DOI:10.3886/ICPSR25383.v2.

United States Department of Health and Human Services. 2011. Substance Abuse and Mental Health Services Administration. Center for Behavioral Health Statistics and Quality. National Survey on Drug Use and Health, 2011. ICPSR34481-v2. Ann Arbor, MI: Inter-university Consortium for Political and Social Research [distributor], 2013-06-20.

Van der Heijden, Peter G. M., Antoine de Falguerolles, and Jan de Leeuw. 1989. "A Combined Approach to Contingency Table Analysis Using Correspondence Analysis and Log-linear Analysis." *Applied Statistics*, 38: 249–292.

Van der Heijden, Peter G. M., and Jan de Leeuw. 1985. "Correspondence Analysis Used Complementary to Loglinear Analysis." *Psychometrika*, 50: 429–447.

6

Binary Logistic Regression

Up to this point the discussion of modeling techniques has focused on general and logit log-linear analysis. The general log-linear model examines associations among categorical variables in a symmetrical manner, while the asymmetrical logit model draws distinctions between explanatory and response measures. The text now turns to logistic regression analysis, a technique that, unlike the contingency-table approach, allows interval-level explanatory measures to be modeled as main effects.[1] Logistic regression models containing continuous explanatory variables analyze data at the individual level (DeMaris 1992), thus resembling ordinary least squares (OLS) regression, which fits a straight line through individual observations. With the logistic model, the OLS straight line becomes an S-shaped sigmoid, and maximum likelihood replaces least squares as an estimation technique (Hosmer and Lemeshow 2000, Kleinbaum 1992, Menard 2002; see, earlier, Cox 1970, Theil 1970).

The present chapter focuses on binary logistic regression, a technique used when a dependent variable contains two categories. Chapter 7 addresses multinomial logistic regression, a procedure applied when a response measure contains more than two nominal categories or when the procedure addressed in Chapter 8, ordinal logistic regression, does not meet statistical assumptions. As its name implies, ordinal regression examines the effects of one or more explanatory variables on an ordered response measure, and the model assumes consistent effects across categories. But before addressing models containing polytomous response measures, the text necessarily covers the fundamentals of binary logistic regression. In doing so, the current chapter uses examples based on the horse-racing data as well as the 2008 American National Election Study and the 2012 Monitoring the Future study. The chapter begins with a review of communication studies that have used logistic regression to test variable relationships.

Examples of Published Research

Of the modeling techniques addressed in this text, communication scholars have used binary logistic regression the most frequently. Most of the research has examined issues in health communication, with scholarship also focusing on research methods, media content and effects, politics and policy matters, and interpersonal communication. Beginning with studies addressing behavioral intentions in health and politics, Agarwal (2011) analyzed factors associated with whether women intended to schedule mammograms (see also Harada et al. 2013), while Avery and Lariscy (2014) analyzed whether individuals planned to receive flu vaccinations. Kim and Shanahan (2003) examined predictors of whether individuals intended to quit smoking and, in political research, Kleinnijenhuis et al. (2007) studied voting intentions. Researchers in health communication have used logistic regression in studies of information seeking (Dobransky and Hargittai 2012, Dunleavy, Crandall, and Metsch 2005, Kim and Kwon 2010, Niederdeppe, Frosch, and Hornik 2008), and in analyses of information disclosure (Aldeis and Afifi 2013, Darst et al. 2014, Koskan et al. 2014). Scholars have examined advertising recall and attitudes toward media health campaigns (Audrain-McGovern et al. 2003, Berry et al. 2011, Davis et al. 2013, Dunlop, Wakefield, and Kashima 2008, Faulkner et al. 2011, Kwak, Fox, and Zinkhan 2002) as well as communication problems occurring during public-health disasters (Taylor-Clark, Viswanath, and Blendon 2010).

In the context of research methods, scholars have used logistic regression in studies of media exposure (Collins 2008), measurement of political participation (Dylko 2010), health-literacy assessment (Haun et al. 2012), and predictive validity in standardized testing (Feeley, Williams, and Wise 2005). In gaming research, Chang, Lee, and Kim (2006) examined factors associated with online game adoption, while Beullens and Van den Bulck (2013) studied whether driving-related game participation predicted youth involvement in car accidents. Beullens, Roe, and Van den Bulck (2013) investigated associations between driving-game participation and driving without a license. Fairlie et al. (2010) examined more general predictors of impaired driving (see also Kenney, LaBrie, and Lac 2013), while Miller et al. (2006) examined predictors of tobacco use. Relatedly, Dunlop, Cotter, and Perez (2014) studied associations among antismoking advertising, interpersonal pressure, and outcomes related to smoking cessation.

In studies of media effects, Dixon et al. (2014) investigated public attitudes about tanning and skin cancer; Witting et al. (2012) studied the influence of message framing on parental decisions to participate in ultrasound screening for developmental hip dysplasia; and Young (2006) examined the effects of late-night television exposure on whether interviewees mentioned caricatured candidate traits. Martin and Wilson (2011) analyzed parental communication about kidnapping reports and, in a content analysis, Lee and Joo (2005) studied the portrayal of Asians in magazine advertisements. In studies involving interpersonal communication, Meyer and Rothenberg (2004) examined message repair strategies, Morgan (2004) investigated family communication about organ donation, and Moorman (2011) studied factors associated with end-of-life health care. Deutsch, Frese, and Sandholzer (2014) also examined issues in family communication, while Mello et al. (2013) used logistic regression to study whether cancer survivors had experienced anxiety or depression during a 12-month period (see also Robinson and Tian 2009). Finally, in the context of legal communication, Napoli (2000) used logistic regression in a study investigating pro-regulatory versus deregulatory broadcast policies.

As this overview of published research demonstrates, communication scholars have used binary logistic regression in multiple contexts. The model accommodates categorical and continuous explanatory measures, making it more flexible than the logit log-linear model. The following section reviews the fundamental components of binary logistic regression, comparing and contrasting it with ordinary least squares regression.

Binary Logistic Regression: Fundamentals

Communication scholars frequently use ordinary least squares regression to analyze the effects of dichotomous and continuous explanatory measures on the behavior of a continuous response variable. The procedure uses a global F test to determine whether a given model fits the data in a statistically acceptable manner. An OLS model identifies significant predictors as well as the percentage of system variance the predictors explain. Statistically, the OLS model is expressed as:

$$Y = \beta_o + \beta_1 X_1 + \beta_2 X_2 + \cdots + \beta_k X_k + \varepsilon$$

where each β estimate represents a change in Y, the dependent variable, when a value for X, an independent measure, increases by one unit. The intercept term β_o represents the value of Y when the value of X equals 0, and error ε represents the arithmetic difference in what a model predicts and the dependent variable. Social scientists have used OLS regression in analyses containing binary response measures, but as Aldrich and Nelson (1984, 14) explained, "even in the best of circumstances, OLS regression estimates of a dichotomous dependent variable are, although unbiased, not very desirable." A review of model assumptions helps to elucidate differences between OLS and binary logistic regression, offering insight on why the latter is preferable for analyses containing dichotomous dependent variables.

Menard (2002, 4–5) discussed eight assumptions of OLS regression, beginning with measurement expectations. Independent variables should be dichotomous, interval, or ratio measures, and a dependent variable should be measured at the interval or ratio level. Regarding model specification, all relevant explanatory measures should be included with irrelevant measures excluded. The mean of the errors should be 0, and error variance should be consistent across independent variables. Menard noted that error variance should be distributed normally with no correlation among error terms produced at different levels of the explanatory measures. Finally, error terms should not correlate with independent variables, and multicollinearity should be absent (see also Berry 1993).

In the social sciences, even the most careful studies may not meet all eight assumptions, but the logistic regression model has assisted scholars in reducing problems associated with measurement expectations. As Hosmer and Lemeshow (2000) explained, regression models, in general, are concerned with the expected value of a dependent measure given the value of an independent variable. The response-measure quantity is termed the *conditional mean*. In an OLS model, the conditional mean of an interval or ratio response variable is not bound in its value, but, as indicated, when a dependent variable takes on two levels (0, 1) – and two levels only – it becomes inherently nonlinear. In such a case, the conditional mean $\pi(X)$ is the *proportion* of cases scoring 1 on the dependent variable; an OLS model (i.e., a linear probability model) cannot ensure such a value, but a curvilinear logistic regression model can (see Agresti 1990, 84–85, Aldrich and Nelson 1984, Menard 2002).[2] Logistic regression estimates the probability of a dependent variable showing a success (or occurrence) relative to a non-success (or non-occurrence) as a function of one (or more) explanatory variable(s). The model is expressed as:

$$\pi\left(X\right) = \frac{e^{\beta_0 \beta_1 X}}{1 + e^{\beta_0 + \beta_1 X}}$$

Where OLS models contain an identity link and assume a standard normal error distribution, logistic regression models use a logit (or probit) link and a binomial distribution.[3] The logit $g(X)$ is linear in its parameters (DeMaris 1992, Hosmer and Lemeshow 2000, 6, Pampel 2000) and allows researchers using the logistic regression model to conceptualize studies in a manner similar to those using ordinary least squares. The logit, or log odds, is expressed as:

$$g\left(X\right) = \ln\left[\frac{\pi\left(X\right)}{1 - \pi\left(X\right)}\right] = \beta_0 + \beta_1 X$$

"It is important to understand that the probability, the odds, and the logit are three different ways to express exactly the same thing," Menard (2002, 13) explained. "Of the three measures, the probability or the odds is probably the most easily understood. Mathematically, however, the logit form of the probability best helps us to analyze dichotomous dependent variables."

Simple Logistic Regression Analysis

To demonstrate how logistic regression generates parameter estimates, one might consider the data shown in Table 6.1. This data, extracted from the horse-racing study (Denham 2014), shows drug-use references in the *Albuquerque Journal* before and after *The New York Times* began its investigative series. In its initial report, *The New York Times* focused heavily on problems occurring in quarter horse racing in New Mexico, and the horse-racing study therefore examined content in the largest newspaper in that state. As indicated in Table 6.1, just 2 (6.5%) of 31 articles published prior to the first investigative report mentioned equine drug use, compared to 20 (55.6%) of 36 articles published after the initial report.

A simple logistic regression analysis containing time period as the explanatory measure and drug-use mention as a response variable produced parameter estimates shown in Table 6.2. The table shows a parameter estimate of 2.897, which exponentiates to 18.125, the odds ratio. Recalling that logistic regression estimates the probability of an observation taking on the higher of two values (i.e., the value 1 in a binary model containing scores 0 and 1), one would conclude that the odds of a report published in the second period containing a drug mention were 18.125 the same odds for an article published in the first period. Returning to Table 6.1, if one reversed the two row categories, such that period two appeared in the top row, a cross-product calculation of $(20)(29)/(16)(2) = 580/32 = 18.125$ confirms the odds ratio.

Next, the Wald test, a value for which is indicated in Table 6.2, is analogous to an OLS *t*-test in that it examines the effect an individual parameter on the dependent variable (Azen and Walker 2011, 189). The null hypothesis anticipates

Table 6.1   Cross-tabulation of time period by drug-use mentions in horse-racing reports

	Equine Drug-Use Mentions in the Albuquerque Journal		
Time Frame	*Mention*	*No Mention*	*Totals*
Before First Investigative Report	*a* 2 (6.5%)	*b* 29 (93.5%)	31
After First Investigative Report	*c* 20 (55.6%)	*d* 16 (44.4%)	36
Totals	22	45	67

Table 6.2   Logistic regression model testing time period as determinant of drug-use mentions in *Albuquerque Journal*

Variable	B	SE	Wald	df	Significance	Exp(B)
Period	2.897	.804	12.975	1	.000	18.125
Constant	−.223	.335	.443	1	.506	.800

no relationship between an explanatory measure and a response variable, which suggests, statistically, that a parameter estimate will not differ from 0. Once calculated, a Wald value is compared to a chi-square distribution value, and if it exceeds the table value, the Wald result is considered statistically significant. The test takes the following form:

$$\left(\frac{\hat{\beta} - \beta}{s_\beta} \right)^2$$

where the value 0 is subtracted from a given parameter estimate, the resulting value is divided by the standard error, and the quantity is then squared. For the values in Table 6.2, the following equation would be used to calculate a Wald value:

$$\left(\frac{2.897 - 0}{.804} \right)^2 = (3.603)^2 = 12.98$$

As Table 6.2 indicates, this value is statistically significant, and the odds ratio of 18.125 offers a measure of association. Substantively, one would conclude that the *Albuquerque Journal* included the drug-use attribute in significantly more news reports following the initial investigative report published in *The New York Times*.

Multiple Logistic Regression Analysis

In most instances, researchers seek to examine more than the effect of a single explanatory measure on a given response variable. Drawing on data gathered in the 2008 American National Election Study (The American National Election Studies 2008) ($N=993$), the chapter now addresses multiple logistic regression analysis. Independent variables include the categorical measures sex, race, and party identification, and continuous variables include age and exposure to radio news. Operationally, the three-category race measure included Whites, Blacks, and members of other races, while a four-category party ID variable included

Democrats, Republicans, Independents, and members of other political parties. Age was measured in years, and radio exposure was measured by the number of days a respondent listened to radio news during the previous week. A dichotomous dependent variable focused on attitudes toward economic regulation. Specifically, ANES respondents indicated whether they believed a strong government presence is necessary for handling complex economic matters ($N=746$) or whether the free market should be allowed to operate without government intervention ($N=247$).

In building OLS regression models, researchers often begin with demographic variables such as sex, race, and age. Entering those variables first requires other independent variables to explain system variance beyond the fundamental characteristics of research participants. As Table 6.3 illustrates, binary logistic regression models accommodate the OLS approach for both categorical (sex, race) and continuous (age) explanatory measures. For categorical predictors, the logistic model reports an overall Wald score, as well as individual estimates and standard errors, for measures containing more than two options; an overall value does not appear for dichotomous predictor variables. For instance, in Table 6.3, sex is a dichotomous variable, and it therefore does not show an overall Wald statistic. Race, however, contains three categories, and it includes an overall significance test, as well as tests for individual estimates. Age, a continuous measure, contains just one parameter estimate.

Regarding effects, a positive parameter estimate increases the odds on the dependent variable (i.e., makes observations in the higher of two values, 0 and 1, more likely), while a negative estimate decreases the odds. As an example, the first demographic variable entered in Table 6.3, sex, showed significance as an independent measure. Specifically, its parameter estimate of .573 exponentiated to an odds ratio of 1.774, indicating that the odds of males showing support for free markets were approximately 1.774 times the same odds for females (the reference category). The opposite pattern emerged for Black respondents and individuals from other races relative to Whites; the odds of Black respondents

Table 6.3 Logistic regression model testing sex, race, and age as determinants of economic attitudes

Variable	*B*	*SE*	*Wald*	*df*	*Sig.*	*Exp(B)*
Sex (Males)	.573	.152	14.209	1	.000	1.774
Race			37.906	2	.000	
Black	−1.189	.207	32.969	1	.000	.304
Other Race	−.810	.268	9.105	1	.003	.445
Age	.010	.004	5.544	1	.019	1.010
Constant	−2.189	.246	78.877	1	.000	.112

indicating support for free markets were just .304 times the same odds for Whites (selected as the reference category), and the odds of members of other races supporting free markets were just .445. Had the overall Wald test not shown significance, interpretation of these parameter estimates would not have been appropriate. Lastly, the continuous variable, age, showed significance, with older respondents slightly more inclined to support free markets.

Depending on the objectives of a study, researchers may choose to enter variables in "blocks," examining whether the addition of one or more variables in a second block improves a model containing variables entered initially, whether a third block of measures offers an improvement on the second, and so forth. Table 6.4 contains variables from Table 6.3 but also includes a second block containing a variable indicating political party identification. As indicated in Table 6.4, party ID showed significance as an explanatory measure, with Democrats less likely to support the free market than Republicans. The table also indicates that Independents did not differ from the reference category, other party. Inclusion of the party ID variable resulted in 'other race' losing significance as a parameter estimate, while sex and age retained significance as predictors. Relative to the statistics in Table 6.3, sex showed a slightly lower Wald score and age showed a slightly higher value.

To illustrate a model containing three blocks, Table 6.5 contains the continuous measure exposure to radio news, adding to the variables shown in Table 6.4. As indicated by the positive parameter estimate, an increase in days per week listening to radio news increased the odds of supporting free markets. Other explanatory variables continued to show significance as well, ultimately allowing a researcher to conclude that each of the five explanatory measures predicted

Table 6.4   Logistic regression model testing sex, race, and age, as well as political party identification, as determinants of economic attitudes

Variable	B	SE	Wald	df	Sig.	Exp(B)
Block 1						
Sex (Males)	.491	.158	9.611	1	.002	1.634
Race			8.199	2	.017	
Black	−.591	.228	6.725	1	.010	.554
Other Race	−.476	.279	2.910	1	.088	.621
Age	.012	.005	6.687	1	.010	1.012
Block 2						
Party ID			55.485	3	.000	
Democrat	−.940	.316	8.855	1	.003	.391
Republican	.723	.312	5.374	1	.020	2.061
Independent	−.315	.302	1.091	1	.296	.730
Constant	−1.951	.256	58.148	1	.000	.142

Table 6.5 Logistic regression model testing sex, race, and age, as well as political party identification and exposure to radio news, as determinants of economic attitudes

Variable	B	SE	Wald	df	Sig.	Exp(B)
Block 1						
Sex (Males)	.448	.160	7.854	1	.005	1.565
Race			8.123	2	.017	
Black	−.592	.229	6.704	1	.010	.553
Other Race	−.472	.280	2.838	1	.092	.624
Age	.013	.005	7.278	1	.007	1.013
Block 2						
Party ID			50.661	3	.000	
Democrat	−.939	.318	8.740	1	.003	.391
Republican	.669	.314	4.535	1	.033	1.953
Independent	−.302	.304	.988	1	.320	.740
Block 3						
Radio News Exp.	.083	.029	8.027	1	.005	1.086
Constant	−2.163	.270	64.231	1	.000	.115

attitudes toward economic regulation. Had it made conceptual sense, the researcher could have entered additional variables in any of the regression blocks.

Interactions

In some cases, the relationship between an independent and a dependent variable differs based on the level of a third measure (i.e., a moderating variable). Jaccard (2001) has discussed interactions for both categorical and continuous predictors in logistic regression analysis, focusing on hierarchically well-formulated (HWF) models; such models contain not only interactive terms but also lower-order terms on which interactions are based (see also Huang and Shields 2000). As an example, Table 6.5 may have included an interaction between sex and race, provided the researcher had a theoretical reason for testing its effects on the response measure. Perhaps White males, in particular, could be expected to support a free market more than other respondents (in this analysis they did not). The interaction term would have been inserted just below the sex and race main-effect variables in the first block.

Statistical software packages allow researchers to create interaction terms quickly and efficiently. Stokes, Davis, and Koch (2012, 221–227) offer instruction for SAS and Norusis (2005) discusses how to incorporate interactions using SPSS programs. In general, interactions can improve logistic regression models, but like main-effect measures, their inclusion should be based on sound theory.

Model Assessment

Just as the Wald test in logistic regression functions as an analog to the OLS t-test, the *likelihood ratio chi-squared test statistic* (also called *model chi-squared*) serves as an analog to the global F statistic (DeMaris 1992, 47).[4] In a logistic regression model, the null hypothesis suggests that all parameter estimates will equal 0 (statistically), and the alternative hypothesis suggests that at least one estimate will not. An initial log-likelihood value, D_o, is based on a model containing only the intercept, while a model log-likelihood value, D_m, contains all predictors (notation from Menard 2002).

To determine whether a given model fits better than one containing only the intercept, one first subtracts D_m from D_o, establishing a model chi-squared value. Because such a calculation results in a negative number, statisticians and statistical software packages multiply both D_o and D_m by –2; thus, the equation $-2 \log(D_o) - [-2 \log(D_m)]$ yields a value for model chi-squared. The value is then compared to a chi-square distribution table based on available degrees of freedom (i.e., the total number of parameters, intercept excluded). Table 6.6 contains log-likelihood estimates for the three ANES models. Examining Table 6.6, the first block shows a –2 log-likelihood score of 1047.487 and a model chi-squared value of 66.556 with 4 degrees of freedom. The model is significantly different from a model containing only the intercept. Block 2 shows a –2 log-likelihood score of 989.639 and a model chi-squared value of 124.404 with 7 degrees of freedom. Because the difference between the first and second models, 57.848 with 3 degrees of freedom, showed significance, the second model is statistically preferable. The five-variable model represented in Block 3 also showed significance, and therefore a researcher would be justified in selecting that model to report as a final representation of variable relationships.

Additional Statistics

Up to this point the chapter has addressed parameter estimates and standard errors, Wald statistics, log-likelihood measures, and values for model chi-squared. Two additional statistics, Nagelkerke as well as Cox and Snell pseudo-R^2 measures,

Table 6.6 Log-likelihood estimates for three binary logistic regression models

Block	–2LL	Model Chi-Square	df	Sig.	Difference in Models
1	1047.487	66.556	4	.001	
2	989.639	124.404	7	.001	57.848 (3 df, $p < .001$)
3	981.699	132.344	8	.001	7.940 (1 df, $p < .01$)

and the Hosmer and Lemeshow goodness-of-fit test, also offer insight into statistical models and variable relationships. In OLS regression models, R^2 indicates an amount of system variance explained by the independent variables, and that is essentially the case with logistic regression; however, some important differences exist between OLS and logistic models. For instance, Allen and Le (2008) noted that R^2 analogs are not invariant to base rates and therefore models cannot be compared across differing data sets. Consequently, authors who use logistic regression must be careful in noting the limitations of R^2 analogs, perhaps reporting that while the Cox and Snell as well as Nagelkerke measures offer general estimations of explained variance, the statistics should not be compared with one another across studies (see, for additional discussion, Denk and Finkel 1992, Hagle and Mitchell 1992).[5]

In the current chapter, –2 log-likelihood values and associated chi-square statistics have shown whether models containing certain blocks of variables offered improvements on models containing the intercept only as well as models containing fewer measures. Technically, the model chi-squared statistic is not a goodness-of-fit test, even though researchers effectively use it as one. In binary logistic regression, the Hosmer and Lemeshow goodness-of-fit test (see Hosmer and Lemeshow 2000) indicates whether specific models fit the data; as with G^2 in log-linear modeling, a non-significant p-value indicates an acceptable fit. The calculations for the test are somewhat complex, and statisticians have identified limitations for nearly all goodness-of-fit tests associated with logistic regression (see Allison 2013, Simonoff 1998). Still, the Hosmer and Lemeshow test can offer researchers information about the extent to which statistical models fit the data. As the current chapter shows, the test is available in SPSS.

Diagnostic Considerations

Commenting on problems that may occur in logistic regression modeling, Menard (2002, 67) followed up on initial work from Pregibon (1981) as well as Hosmer and Lemeshow (2000) in identifying *bias, inefficiency*, and *invalid statistical inference* as points of concern. Bias refers to systematic misestimations, inefficiency refers to large standard errors relative to corresponding parameter estimates, and invalid statistical inference reflects inaccurate significance testing. Menard also mentioned *high-leverage* instances, where independent variables may contain unusually high or low values, such that they become influential cases in regression equations.

As with OLS regression, Menard (2002, 67) explained, correct model specification is a key assumption in the binary logistic model. When analyses include irrelevant variables, a model may become relatively inefficient, and when researchers eliminate relevant variables, parameter estimates may become biased.

As Menard noted, problems with model specification often occur because the theoretical framework guiding a study has not been adequately developed.

The assumption of linearity in the logit transformation is also an important consideration. A change in the log of the odds should reflect a consistent one-unit change in the independent variable. When linearity is in question, statisticians frequently use a Box-Tidwell test (see Box and Tidwell 1962, Box and Cox 1964) to detect problems. With this approach, a researcher creates a variable containing an explanatory measure multiplied by its natural log and enters the resulting measure in a regression equation. If the variable shows significance, linearity may be lacking (see Menard 2002, 70).

In OLS regression models, problems arise when explanatory measures correlate with one another, and the same is true for logistic regression. Menard (2002) explained that while low levels of *collinearity* may not compromise a logistic regression model, variables that correlate at .8 or higher will almost certainly increase standard errors. In statistical software programs, logistic regression techniques typically lack a formal test for collinearity; to address this problem, Menard advised entering explanatory and response measures into an OLS model and examining tolerance statistics. Functional forms of variable relationships, he noted, are not relevant to the detection of highly correlated predictors. In other words, one can use OLS models to test for collinearity.

Researchers should also consider influential cases in data analyses. One approach to gauging influential cases is to examine Studentized residuals in logistic regression output. Menard (2002, 84) noted that values less than –2 or greater than +2 may indicate poor fit. Most statistical software programs also allow researchers to assess outliers and influential cases using charts and graphs. But before eliminating cases that appear to offer undue influence, researchers should consider whether such cases appear for a substantive reason.

Lastly, scholars who work with large datasets containing rare events should review scholarship that addresses potential problems in regression equations (see, for discussion, Allison 2012). King and Zeng (2001) observed that logistic regression analysis can substantially underestimate the probability of rare events, also explaining that data collection methods can cause problems. Researchers who work with large public datasets that contain rare occurrences would want to consider strategies advised by King and Zeng (2001) before analyzing data. The chapter now addresses logistic regression analysis in SPSS, drawing on data gathered in the 2012 Monitoring the Future study (Johnston et al. 2012).

Binary Logistic Regression in SPSS

Health communication scholars often study patterns of substance use among adolescents, and the current chapter draws on data gathered in the 2012 Monitoring the Future study ($N = 1,775$) to examine sex, race, drug-spot exposure, and frequency of alcohol consumption as determinants of narcotic use. Separate statistical models include an index measuring sensation-seeking, also an important construct in health research. Operationally, the MTF survey asked respondents to indicate the frequency with which they had seen anti-drug spots on television, with categories including not at all, once a month, 1 to 3 times a month, 1 to 3 times a week, and daily or almost daily. The survey also asked respondents to indicate the number of times they had consumed alcohol, beyond a few sips, in their lives. Response intervals included the following: 0, 1–2, 3–5, 6–9, 10–19, 20–39, and 40 or more times. The MTF data included a three-category race measure (Black, White, Hispanic), and for purposes of illustration, a collapsed dependent variable indicated whether respondents had ever tried a narcotic substance.

Figure 6.1 displays two logistic regression windows in SPSS. On the left is the first window a researcher encounters after selecting Analyze > Regression > Binary Logistic. The window allows the researcher to enter a dichotomous dependent variable as well as multiple independent measures. In analyses containing categorical predictors, a researcher must define them as such prior to fitting statistical models. Looking to the window on the right in Figure 6.1, this step involves defining dichotomous measures as indicator (i.e., "dummy") variables and multi-category predictors as simple categorical measures. For the latter, the researcher can establish as a reference category the first or the last variable option. SPSS considers categorical predictors as indicators by default, and a researcher therefore must select an alternative when necessary, clicking Change when complete.

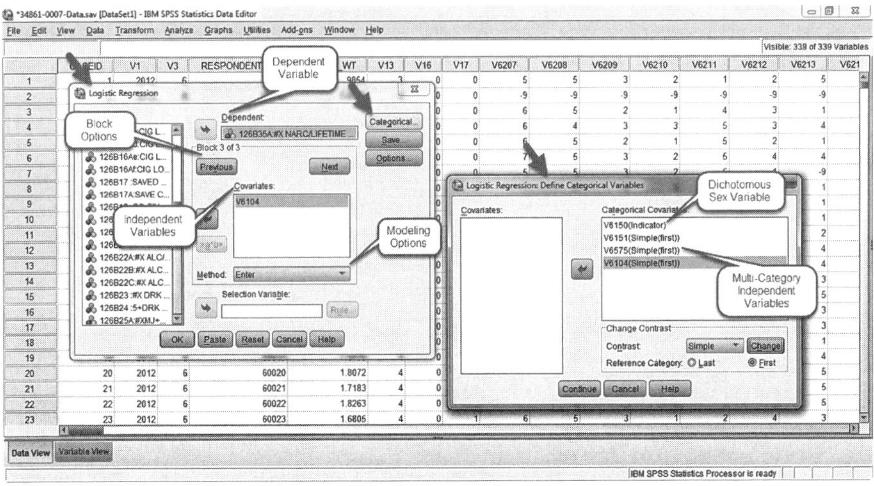

Figure 6.1   SPSS windows for binary logistic regression and variable definition. Source: SPSS® Reprints Courtesy of International Business Machines Corporation, © 2014 International Business Machines Corporation

In the current example, Black respondents, those who had not seen anti-drug spots, and those who had not consumed alcohol serve as reference categories.

In addition to selecting independent and dependent variables, and defining categorical measures, researchers need to select a modeling strategy. The first window in Figure 6.1 contains a drop-down menu, next to Method. This menu allows the researcher to choose an approach for entering explanatory measures, just as in OLS regression analysis. One can choose backward, forward, and stepwise approaches, among others. Additionally, one can choose to enter variables in blocks to fit a research design. The current example uses that approach for purposes of demonstration.

Figure 6.2 contains the main window for binary logistic regression as well as an Options window on the right side. In the Options window, the researcher can (and should) select the Hosmer-Lemeshow goodness-of-fit test, a 95% confidence interval for each exponentiated parameter estimate, and a correlation matrix for variables in the equation. Each of these items can prove useful for evaluating variable relationships and possible instances of collinearity as well as overall fit. In analyses that might contain influential cases, researchers can select a casewise listing of residuals; doing so is especially important when working with small samples.

Table 6.7 contains output corresponding to the menu selections shown in Figures 6.1 and 6.2. Starting at the top of the table, the Dependent Variable Coding box shows internal values for the binary outcome measure. As shown, 0 denoted no narcotic use and 1 denoted one or more instances of use.

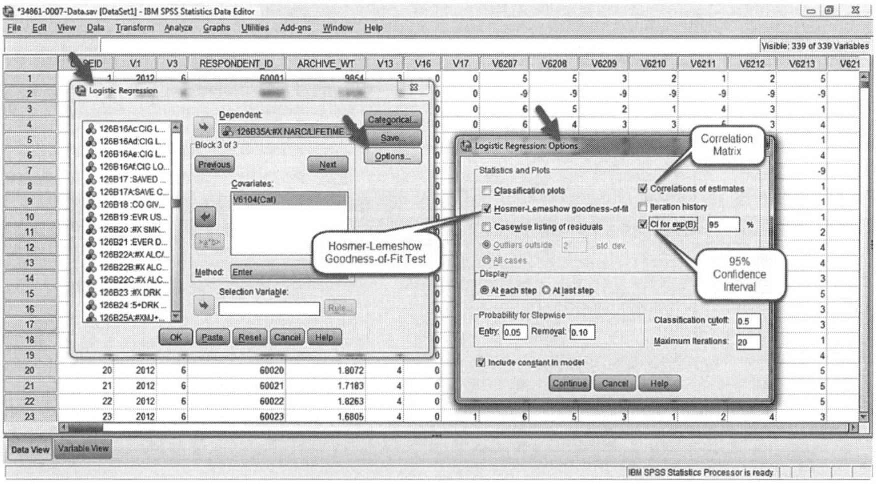

Figure 6.2 SPSS windows for binary logistic regression and analysis options. Source: SPSS® Reprints Courtesy of International Business Machines Corporation, © 2014 International Business Machines Corporation

Table 6.7 SPSS output for binary logistic regression model containing categorical predictors

Dependent Variable Coding

Original Value	Internal Value
0 OCCAS: (1)	0
1+ (2)	1

Categorical Variable Codings

			Frequency	Parameter Coding (1)	(2)	(3)	(4)	(5)	(6)
126B22A:	0 OCCAS:	(1)	509	–.143	–.143	–.143	–.143	–.143	–.143
#X ALC/	1–2X:	(2)	128	.857	–.143	–.143	–.143	–.143	–.143
LIF SIPS	3–5X:	(3)	156	–.143	.857	–.143	–.143	–.143	–.143
	6–9X:	(4)	167	–.143	–.143	.857	–.143	–.143	–.143
	10–19X:	(5)	248	–.143	–.143	–.143	.857	–.143	–.143
	20–39X:	(6)	235	–.143	–.143	–.143	–.143	.857	–.143
	40+OCCAS:	(7)	332	–.143	–.143	–.143	–.143	–.143	.857
126D08:	NOT@ALL:	(1)	454	–.200	–.200	–.200	–.200		
#X SEE	<ONCE/MO:	(2)	428	.800	–.200	–.200	–.200		
DRG SPTS	1–3/MO:	(3)	476	–.200	.800	–.200	–.200		
	1–3/WK:	(4)	281	–.200	–.200	.800	–.200		
	DAILY:	(5)	136	–.200	–.200	–.200	.800		
126C04®:	BLACK:	(1)	170	.000	.000				
Rs RACE	WHITE:	(2)	1332	1.000	.000				
B/W/H	HISPANIC:	(3)	273	.000	1.000				
126C03:	MALE:	(1)	892	1.000					
Rs SEX	FEMALE:	(2)	883	.000					

Block 0: Beginning Block

Classification Table[a,b]

			Predicted		
			126B35A: #X NARC/ LIFETIME		Percentage
Observed			0 OCCAS: (1)	1+: (2)	Correct
Step 0	126B35A: #X	0 OCCAS: (1)	1561	0	100.0
	NARC/LIFETIME	1+: (2)	214	0	.0
	Overall Percentage				87.9

[a] Constant included in the model.
[b] The cut value is .500

(*Continued*)

Table 6.7 (Continued)

Variables in the Equation

		B	SE	Wald	df	Sig.	Exp(B)
Step 0	Constant	−1.987	.073	743.122	1	.000	.137

Variables Not in the Equation

			Score	df	Sig.
Step 0	Variables	Sex (Males)	5.757	1	.016
		Race	10.637	2	.005
		White	6.737	1	.009
		Hispanic	.347	1	.556
	Overall Statistics		14.920	3	.002

Block 1: Method = Enter

Omnibus Tests of Model Coefficients

			Chi-Square	df	Sig.
Step 1		Step	17.199	3	.001
		Block	17.199	3	.001
		Model	17.199	3	.001

Model Summary

Step	−2 Log Likelihood	Cox & Snell R Square	Nagelkerke R Square
1	1289.364[a]	.010	.019

[a] Estimation terminated at iteration number 5 because parameter estimates changed by less than .001.

Hosmer and Lemeshow Test

Step	Chi-Square	df	Sig.
1	1.271	3	.736

Contingency Table for Hosmer and Lemeshow Test

		126B35A: #X NARC/LIFETIME = 0 OCCAS		126B35A: #X NARC/LIFETIME = 1+ OCCAS		
		Observed	Expected	Observed	Expected	Total
Step 1	1	162	162.000	8	8.000	170
	2	129	126.652	11	13.348	140
	3	556	560.023	76	71.977	632
	4	114	116.348	19	16.652	133
	5	600	595.977	100	104.023	700

Table 6.7 (Continued)

Classification Table[a]

	Observed		Predicted		
			126B35A: #X NARC/ LIFETIME		*Percentage*
			0 OCCAS: (1)	*1+: (2)*	*Correct*
Step 1	126B35A: #X	0 OCCAS: (1)	1561	0	100.0
	NARC/LIFETIME	1+: (2)	214	0	.0
	Overall Percentage				87.9

[a]The cut value is .500

Variables in the Equation

		B	SE	Wald	df	Sig.	Exp(B)	95% CI for Exp(B) Lower	Upper
Step 1[a]	V6150(1)	.306	.148	4.264	1	.039	1.358	1.016	1.816
	V6151			8.797	2	.012			
	V6151(1)	1.073	.372	8.306	1	.004	2.923	1.409	6.062
	V6151(2)	.874	.411	4.515	1	.034	2.397	1.070	5.368
	Constant	−3.124	.367	72.282	1	.000	.044		

Block 2: Method = Enter

Omnibus Tests of Model Coefficients

		Chi-Square	df	Sig.
Step 1	Step	1.950	4	.745
	Block	1.950	4	.745
	Model	19.149	7	.008

Model Summary

Step	−2 Log Likelihood	Cox & Snell R Square	Nagelkerke R Square
1	1287.414[a]	.011	.021

[a]Estimation terminated at iteration number 6 because parameter estimates changed by less than .001.

Hosmer and Lemeshow Test

Step	Chi-Square	df	Sig.
1	2.010	8	.981

(*Continued*)

Table 6.7 (Continued)

Contingency Table for Hosmer and Lemeshow Test

		126B35A: #X NARC/ LIFETIME = 0 OCCAS		126B35A: #X NARC/ LIFETIME = 1+ OCCAS		
		Observed	Expected	Observed	Expected	Total
Step 1	1	162	162.000	8	8.000	170
	2	206	205.274	21	21.726	227
	3	97	97.549	12	11.451	109
	4	170	168.045	19	20.955	189
	5	104	107.068	18	14.932	122
	6	164	164.639	24	23.361	188
	7	149	148.750	22	22.250	171
	8	203	206.245	37	33.755	240
	9	145	141.278	23	26.722	168
	10	161	160.152	30	30.848	191

Classification Table[a]

			Predicted		
			126B35A: #X NARC/ LIFETIME		Percentage
	Observed		0 OCCAS: (1)	1+: (2)	Correct
Step 1	126B35A: #X	0 OCCAS: (1)	1561	0	100.0
	NARC/LIFETIME	1+: (2)	214	0	.0
	Overall Percentage				87.9

[a] The cut value is .500

Variables in the Equation

								95% CI for Exp(B)	
		B	SE	Wald	df	Sig.	Exp(B)	Lower	Upper
Step 1[a]	V6150(1)	.306	.149	4.237	1	.040	1.358	1.015	1.816
	V6151			8.467	2	.014			
	V6151(1)	1.059	.375	7.983	1	.005	2.884	1.383	6.013
	V6151(2)	.858	.413	4.321	1	.038	2.358	1.050	5.293
	V6575			1.925	4	.750			
	V6575(1)	.108	.216	.251	1	.617	1.114	.729	1.703
	V6575(2)	.253	.206	1.513	1	.219	1.288	.861	1.927
	V6575(3)	.231	.235	.960	1	.327	1.259	.794	1.998
	V6575(4)	.248	.311	.634	1	.426	1.281	.696	2.356
	Constant	−3.097	.368	70.876	1	.000	.045		

Table 6.7 (Continued)

Block 3: Method = Enter

Omnibus Tests of Model Coefficients

		Chi-Square	df	Sig.
Step 1	Step	201.423	6	.000
	Block	201.423	6	.000
	Model	220.572	13	.000

Model Summary

Step	−2 Log Likelihood	Cox & Snell R Square	Nagelkerke R Square
1	1085.991[a]	.117	.224

[a] Estimation terminated at iteration number 8 because parameter estimates changed by less than .001.

Hosmer and Lemeshow Test

Step	Chi-Square	df	Sig.
1	2.509	8	.961

Contingency Table for Hosmer and Lemeshow Test

		126B35A: #X NARC/ LIFETIME = 0 OCCAS		126B35A: #X NARC/ LIFETIME = 1+ OCCAS		
		Observed	Expected	Observed	Expected	Total
Step 1	1	174	174.513	2	1.487	176
	2	170	170.463	3	2.537	173
	3	180	177.541	1	3.459	181
	4	178	178.192	5	4.808	183
	5	172	171.676	12	12.324	184
	6	158	159.765	22	20.235	180
	7	162	161.176	27	27.824	189
	8	139	137.185	33	34.815	172
	9	117	118.983	48	46.017	165
	10	111	111.507	61	60.493	172

Classification Table[a]

			Predicted		
			126B35A: #X NARC/LIFETIME		Percentage
	Observed		0 OCCAS: (1)	1+: (2)	Correct
Step	126B35A: #X	0 OCCAS: (1)	1561	0	100.0
1	NARC/LIFETIME	1+: (2)	214	0	.0
	Overall Percentage				87.9

[a] The cut value is .500

(Continued)

Table 6.7   (Continued)

								95% CI for Exp(B)	
		B	*SE*	*Wald*	*df*	*Sig.*	*Exp(B)*	*Lower*	*Upper*
Step 1[a]	V6150(1)	.175	.159	1.225	1	.268	1.192	.873	1.626
	V6151			3.210	2	.201			
	V6151(1)	.502	.393	1.632	1	.201	1.652	.765	3.568
	V6151(2)	.198	.433	.209	1	.648	1.219	.522	2.847
	V6575			2.716	4	.606			
	V6575(1)	.095	.229	.172	1	.679	1.100	.702	1.724
	V6575(2)	.302	.217	1.935	1	.164	1.353	.844	2.072
	V6575(3)	.298	.251	1.418	1	.234	1.348	.825	2.202
	V6575(4)	.273	.330	.686	1	.407	1.314	.688	2.510
	V6104			127.230	6	.000			
	V6104(1)	−.787	1.059	.552	1	.458	.455	.057	3.630
	V6104(2)	.847	.536	2.491	1	.114	2.332	.815	6.670
	V6104(3)	1.875	.424	19.534	1	.000	6.521	2.839	14.975
	V6104(4)	2.067	.388	28.382	1	.000	7.897	3.692	16.892
	V6104(5)	2.562	.376	46.351	1	.000	12.967	6.201	27.116
	V6104(6)	3.221	.358	80.916	1	.000	25.056	12.420	50.549
	Constant	−3.118	.406	58.878	1	.000	.044		

With use taking on the higher value, logistic regression models estimated the probability of respondents having experimented at least once with a narcotic substance. The next section, Categorical Variable Codings, shows both frequencies and internal coding for explanatory measures. Following that descriptive information, Block 0, often referred to as the "null hypothesis block," contains output based on a model containing the intercept only. The null hypothesis suggests that explanatory measures will not differ from 0, and the statistics in this block inform that prediction. The Classification Table shows overall predictability, and in this case, 87.9% of cases (an unusually high percentage) could be predicted based on the model. The Variables in the Equation box shows an estimate, standard error, Wald value, and exponentiated estimate for the constant. Lastly, the Variables Not in the Equation box indicates whether measures not included might contribute to subsequent models. Here it appears they will.

In Block 1, the Omnibus Tests of Model Coefficients section indicates whether variables contained in the first block improved the regression model. The model chi-squared value, 17.199 with 3 degrees of freedom, shows significance and indicates improvement; however, as indicated in the Model Summary, sex and race accounted for less than 2% of the variation in the dependent variable.

The Hosmer-Lemeshow Test indicates a good fit to the data, and the Contingency Table for [the] Hosmer-Lemeshow Test shows small differences between observed and expected frequencies. Although the Classification Table shows little change between Blocks 0 and 1, the Variables in the Equation section nevertheless shows significant effects for sex and race. Here, the odds of a male using a narcotic substance were 1.358 times the same odds for a female. Recalling that Black respondents served as the reference group for race, the odds of a White adolescent trying a narcotic substance were nearly three times the odds of a Black respondent doing so, with the odds of an Hispanic respondent using a narcotic substance more than twice those of a Black adolescent.

The next block includes exposure to anti-drug spots on television, and as indicated by the Omnibus Tests of Model Coefficients section, this variable added little to the model. The block was not significant, which the Variables Not in the Equation section suggested. Looking to the Variables in the Equation section, and specifically to the 95% confidence intervals, each category in the drug-spot exposure variable passed through 1.0, indicating independence, or no effect. Thus, with sex and race controlled, exposure to drug-spots did not serve as a significant determinant of narcotic experimentation.

Block 3 contains lifetime alcohol consumption as an explanatory measure, and unlike drug-spot exposure, it shows significance as a predictor. With alcohol consumption in the model, explained variance increased from approximately 2% to 22.4%, and both sex and race became non-significant as predictors. The relationship between alcohol consumption and narcotic use moved in the kind of linear pattern one would expect; that is, the more frequently an individual consumed alcohol, the more likely he or she was to experiment with a narcotic substance. As examples, the odds of those who had consumed alcohol between 10 and 19 times experimenting with a narcotic substance were 7.897 times the same odds for those who had not consumed alcohol. The odds for those who had consumed alcohol between 20 and 39 times were 12.967, and the odds for those who had consumed alcohol on 40 or more occasions were 25.056. Notably, with the drug-spot exposure variable retained in the model, standard errors appeared larger than they would have otherwise.

Table 6.8 displays the explanatory effects of sex and race, as well as a six-item, 30-point index measuring sensation-seeking, on narcotic experimentation. The purpose of this example is to demonstrate how continuous variables may appear in the same regression equation as categorical measures – and how a continuous variable can affect explanatory power. Although the samples between analyses shown in Tables 6.7 and 6.8 showed slight variation, initial results were similar: Block 1, which contained sex and race, offered an improvement on the model containing the intercept only (Block 0). In Block 1, the sex measure did not show statistical significance ($p=.062$), but White and Hispanic adolescents again

Table 6.8 SPSS output for binary logistic regression model containing categorical and continuous predictors

Dependent Variable Coding

Original Value	Internal Value
0 OCCAS: (1)	0
1+ (2)	1

Categorical Variable Codings

			Frequency	Parameter Coding (1)	(2)
126C04[®]: Rs RACE B/W/H	BLACK:	(1)	173	.000	.000
	WHITE:	(2)	1304	1.000	.000
	HISPANIC:	(3)	272	.000	1.000
126C03: Rs SEX	MALE:	(1)	876	1.000	
	FEMALE:	(2)	873	.000	

Block 0: Beginning Block

Classification Table[a,b]

			Predicted 126B35A: #X NARC/ LIFETIME 0 OCCAS: (1)	1+: (2)	Percentage Correct
	Observed				
Step 0	126B35A: #X NARC/LIFETIME	0 OCCAS: (1)	1543	0	100.0
		1+: (2)	206	0	.0
	Overall Percentage				88.2

[a] Constant included in the model.

[b] The cut value is .500

Variables in the Equation

		B	S.E.	Wald	df	Sig.	Exp(B)
Step 0	Constant	−2.014	.074	736.874	1	.000	.134

Variables Not in the Equation

			Score	df	Sig.
Step 0	Variables	Sex (Males)	4.836	1	.028
		Race	12.268	2	.002
		White	7.814	1	.005
		Hispanic	.386	1	.534
	Overall Statistics		15.773	3	.001

Table 6.8   (Continued)

Block 1: Method = Enter

Omnibus Tests of Model Coefficients

		Chi-Square	df	Sig.
Step 1	Step	18.724	3	.001
	Block	18.724	3	.001
	Model	18.724	3	.001

Model Summary

Step	−2 Log Likelihood	Cox & Snell R Square	Nagelkerke R Square
1	1249.237[a]	.011	.021

[a] Estimation terminated at iteration number 5 because parameter estimates changed by less than .001.

Hosmer and Lemeshow Test

Step	Chi-Square	df	Sig.
1	1.314	3	.726

Contingency Table for Hosmer and Lemeshow Test

		126B35A: #X NARC/ LIFETIME = 0 OCCAS		126B35A: #X NARC/ LIFETIME = 1+ OCCAS		
		Observed	Expected	Observed	Expected	Total
Step 1	1	166	166.000	7	7.000	173
	2	124	121.525	10	12.475	134
	3	552	555.561	75	71.439	627
	4	119	121.475	19	16.525	138
	5	582	578.439	95	98.561	677

Classification Table[a]

			Predicted		
			126B35A: #X NARC/ LIFETIME		Percentage
	Observed		0 OCCAS: (1)	1+: (2)	Correct
Step 1	126B35A: #X	0 OCCAS: (1)	1543	0	100.0
	NARC/LIFETIME	1+: (2)	206	0	.0
	Overall Percentage				88.2

[a] The cut value is .500

(*Continued*)

Table 6.8 (Continued)

Variables in the Equation

		B	S.E.	Wald	df	Sig.	Exp(B)	95% C.I. for Exp(B) Lower	Upper
Step 1[a]	V6150(1)	.281	.151	3.496	1	.062	1.325	.987	1.780
	V6151			10.217	2	.006			
	V6151(1)	1.223	.395	9.562	1	.002	3.396	1.565	7.372
	V6151(2)	.997	.434	5.288	1	.021	2.711	1.159	6.345
	Constant	-3.274	.391	70.113	1	.000	.038		

Block 2: Method = Enter

Omnibus Tests of Model Coefficients

		Chi-Square	df	Sig.
Step 1	Step	50.488	1	.000
	Block	50.488	1	.000
	Model	69.212	4	.000

Model Summary

Step	−2 Log Likelihood	Cox & Snell R Square	Nagelkerke R Square
1	1198.749[a]	.039	.075

[a] Estimation terminated at iteration number 6 because parameter estimates changed by less than .001.

Hosmer and Lemeshow Test

Step	Chi-Square	df	Sig.
1	8.089	8	.425

Contingency Table for Hosmer and Lemeshow Test

		126B35A: #X NARC/LIFETIME = 0 OCCAS		126B35A: #X NARC/LIFETIME = 1+ OCCAS		
		Observed	Expected	Observed	Expected	Total
Step 1	1	168	166.215	3	4.785	171
	2	163	165.230	11	8.770	174
	3	157	154.495	9	11.505	166
	4	160	161.165	16	14.835	176
	5	140	138.866	14	15.134	154
	6	161	155.426	14	19.574	175
	7	141	142.687	23	21.313	164
	8	131	134.659	27	23.341	158
	9	140	148.201	40	31.799	180
	10	182	176.056	49	54.944	231

Table 6.8   (Continued)

Classification Table[a]

			Predicted		
			126B35A: #X NARC/ LIFETIME		*Percentage*
	Observed		*0 OCCAS: (1)*	*1+: (2)*	*Correct*
Step 1	126B35A: #X	0 OCCAS: (1)	1543	0	100.0
	NARC/LIFETIME	1+: (2)	206	0	.0
	Overall Percentage				88.2

[a]The cut value is .500

Variables in the Equation

		B	*S.E.*	*Wald*	*df*	*Sig.*	*Exp(B)*	*95% C.I. for Exp(B)* Lower	Upper
Step 1[a]	V6150(1)	.070	.156	.204	1	.651	1.073	.791	1.455
	V6151			8.957	2	.011			
	V6151(1)	1.126	.399	7.980	1	.005	3.083	1.412	6.732
	V6151(2)	.863	.437	3.892	1	.049	2.379	1.006	5.584
	Sensation	.105	.015	45.890	1	.000	1.110	1.077	1.145
	Constant	−5.350	.514	108.211	1	.000	.005		

differed significantly from Black respondents. The odds of a White respondent experimenting with a narcotic substance were 3.396 times the same odds for a Black respondent, with odds for an Hispanic adolescent slightly less, at 2.711.

Looking to Block 2, adding the sensation-seeking index increased the model chi-squared value significantly, from 18.724 with 3 degrees of freedom to 69.212 with 4 degrees of freedom. Including the index also moved the Nagelkerke R-square measure from 2.1% to 7.5%. This addition did not prove as notable as the addition of the alcohol-consumption variable in Table 6.7, but like the alcohol measure, it affected the demographic variables already in the model. The variables in the Equation section for Block 2 indicates that gender moved from a p-value approaching significance ($p = .062$) to .651 with the sensation-seeking index included. Race categories remained significant, although the odds appeared slightly lower. The sensation-seeking index itself did not appear especially notable in terms of odds, but such a value is typical of continuous variables. Although they may not appear important when viewed in isolation, continuous measures do stand to affect explained variation in the dependent variable as well as the individual relationships between explanatory and response measures.

Chapter Summary

This chapter has addressed binary logistic regression, a procedure used to analyze the effects of categorical and continuous explanatory measures on a dichotomous response variable. In terms of generalized linear models, where OLS analyses contain an identity link and assume a standard normal error distribution, logistic regression models use a logit link and assume a binomial distribution. OLS models are based on least squares estimation techniques, while estimates in the logistic regression model are based on maximum likelihood. In addition to binary logistic regression, communication scholars also can apply multinomial and ordinal logistic regression, both of which are covered in the current text.

Chapter Exercises

1. Journalists sometimes grant anonymity to news sources, protecting suppliers of sensitive information from professional and/or personal repercussions. Denham (2012) examined the prevalence of anonymous attribution in *The New York Times*, specifically in reports addressing the convergence of the US war on drugs and the US war on terror. The study also examined other types of sources used in news reports, including those representing the Office of National Drug Control Policy (ONDCP).

    Given the raw data that follow (gathered for the study described above), use binary logistic regression to examine the effects of time period (Before/ After September 11, 2001), article placement on the front page (Yes/No), and dateline (US/International origin) on whether articles contained anonymous attribution. As part of your report, include the –2 log likelihood value, the Cox and Snell R-squared statistic, the Nagelkerke R-squared statistic, and the Hosmer and Lemeshow goodness-of-fit statistic. Additionally, include parameter estimates and their standard errors as well significance levels and 95% confidence intervals. What might be concluded about the effects of the three explanatory measures on the use of anonymous attribution?

2. Use binary logistic regression to examine the effects of time period, article placement on the front page, and dateline on whether articles referenced a source from the Office of National Drug Control Policy. As part of your report, include the –2 log likelihood value, the Cox and Snell R-squared statistic, the Nagelkerke R-squared statistic, and the Hosmer and Lemeshow goodness-of-fit statistic. Additionally, include parameter estimates and their standard errors as well significance levels and 95% confidence intervals.

What might be concluded about the effects of the three explanatory measures on ONDCP sourcing? How did the results of this analysis differ from the results of the analysis of anonymous attribution?

ID	A P F D	O	ID	A P F D	O	ID	A P F D	O	ID	A P F D	O	ID	A P F D	O
001	0 1 2 1	0	034	0 1 2 1	0	067	1 2 2 2	0	100	1 2 2 2	0	133	0 1 2 2	0
002	1 1 1 1	0	035	1 1 2 1	1	068	1 2 2 1	0	101	1 2 2 1	0	134	0 1 2 2	1
003	0 1 2 1	0	036	0 1 2 1	1	069	1 2 2 2	0	102	1 2 2 1	0	135	1 1 1 1	0
004	1 1 1 1	0	037	0 1 2 2	0	070	1 2 2 1	0	103	1 2 2 1	0			
005	0 1 2 1	1	038	1 1 2 1	0	071	1 2 2 1	0	104	1 2 2 1	0			
006	1 1 1 1	1	039	0 1 2 1	0	072	1 2 1 1	0	105	1 1 2 1	0			
007	0 1 2 1	0	040	1 1 2 1	0	073	0 2 2 1	0	106	0 1 2 1	0			
008	1 1 2 2	0	041	0 1 2 1	0	074	1 2 2 2	0	107	0 1 2 1	0			
009	1 1 2 1	0	042	1 1 2 1	0	075	0 2 2 2	0	108	0 1 2 2	0			
010	0 1 2 1	0	043	1 1 2 1	1	076	0 2 2 1	0	109	0 1 2 1	0			
011	1 1 2 2	0	044	1 1 2 2	1	077	1 2 2 1	0	110	0 1 2 1	0			
012	0 1 2 1	0	045	1 1 2 2	1	078	0 2 2 1	0	111	0 1 2 1	0			
013	1 1 2 1	0	046	0 1 1 1	0	079	1 2 2 2	0	112	1 1 2 1	1			
014	1 1 2 2	0	047	1 1 1 1	1	080	1 2 1 1	0	113	1 1 2 2	0			
015	1 1 1 1	1	048	0 1 2 1	1	081	1 2 2 2	0	114	0 1 2 2	1			
016	1 1 1 2	0	049	0 1 2 1	1	082	1 2 2 2	0	115	1 1 1 2	0			
017	1 1 2 1	1	050	1 1 1 1	0	083	1 2 2 1	0	116	1 1 1 1	0			
018	0 1 2 1	0	051	1 1 1 1	1	084	1 2 2 1	0	117	1 1 2 1	0			
019	1 1 1 2	0	052	1 1 2 2	0	085	1 2 1 1	1	118	0 1 2 1	0			
020	1 1 1 2	1	053	0 1 2 1	0	086	1 2 2 1	0	119	1 1 2 1	0			
021	1 1 2 2	0	054	1 2 2 2	0	087	1 2 2 1	0	120	0 1 2 1	1			
022	1 1 1 2	1	055	0 2 2 1	0	088	0 2 2 1	0	121	0 1 2 1	1			
023	1 1 1 2	0	056	0 2 2 1	0	089	1 2 1 1	0	122	0 1 2 1	0			
024	1 1 1 1	1	057	1 2 1 1	0	090	1 2 2 2	0	123	1 1 2 2	0			
025	0 1 2 1	0	058	1 2 1 2	0	091	1 2 2 2	0	124	1 1 2 2	1			
026	0 1 2 1	0	059	1 2 2 2	0	092	1 2 2 2	0	125	1 1 2 2	0			
027	1 2 2 1	0	060	0 2 1 1	0	093	1 2 2 2	1	126	0 1 2 1	1			
028	0 1 2 2	1	061	0 2 2 1	0	094	1 2 2 1	1	127	1 1 2 2	1			
029	1 1 1 2	0	062	0 2 1 2	0	095	0 2 2 2	0	128	0 1 2 1	0			
030	0 1 2 1	0	063	1 2 2 2	0	096	0 2 2 2	0	129	1 1 2 1	0			
031	0 1 2 1	0	064	0 2 2 1	1	097	1 2 1 1	0	130	1 1 2 2	0			
032	0 1 2 2	0	065	1 2 2 2	0	098	0 2 2 1	0	131	0 1 2 1	1			
033	1 1 2 2	0	066	0 2 2 1	0	099	1 2 2 2	0	132	0 1 2 1	0			

Note: Excel file containing data available on companion website.

Category codes:

Anonymous attribution present (A):	0 = No, 1 = Yes
Period (P):	1 = Before 9/11, 2 = After 9/11
Front page news report (F):	1 = Yes, 2 = No
Dateline (D):	1 = US, 2 = International
Office of National Drug Control Policy source present (O):	0 = No, 1 = Yes

Notes

1. As indicated in Chapters 4 and 5, although log-linear models accommodate continuous covariates, they do not apply them on a case-by-case basis. Additional discussion on similarities and differences between log-linear modeling and logistic regression can be found in Imrey, Koch, and Stokes (1981, 1982). Tansey et al. (1996) compared the techniques through scholarship in organizational management.
2. As Menard (2002, 6–7) explained, "Coding the values of the variable as 0 and 1 produces the result that the mean of the variable is the proportion of cases in the higher of the two categories of the variable, and the predicted value for the dependent variable ... can be interpreted as the predicted probability that a case falls into the higher of the two categories ... given its value on the independent variable."
3. Fienberg (2007) discussed the use of discriminant analysis versus logistic regression, noting that, in the former, explanatory variables need to follow the multivariate normal distribution. Logistic regression does not require normally distributed data, and as Fienberg explained, the logistic regression model may actually prove more effective for both classification and prediction. Regarding the logit and probit link functions, the former refers to the log of the odds and the latter to the inverse of the cumulative normal distribution (see, e.g., Allison 1999, Bliss 1935, Caffo and Griswold 2006, Liao 1994).
4. Scholars sometimes consider model chi-squared a measure of goodness of fit, but as DeMaris (1992, 47) explained, it actually is not such an indicator: "It should be emphasized that the model chi-squared test is quite different from G^2, the goodness of fit measure used in contingency-table analyses. Whereas G^2 assesses the fit of the model to the data, the model chi-squared in logistic regression tests only whether any of the predictors are linearly related to the log odds of the event of interest."
5. Richard Williams offers a discussion of R^2 analogs at http://www3.nd.edu/~rwilliam/stats3/L05.pdf.

References

Agarwal, Vinita. 2011. "Investigating the Contribution of Benefits and Barriers on Mammography Intentions of Middle Class Urban Indian Women: An Exploratory Study." *Communication Research Reports*, 28: 235–243. DOI:10.1080/08824096.2011.586254.

Agresti, Alan. 1990. *Categorical Data Analysis.* New York: John Wiley & Sons.

Aldeis, Desiree, and Tamara D. Afifi. 2013. "College Students' Willingness to Reveal Risky Behaviors: The Influence of Relationship and Message Type." *Journal of Family Communication*, 13: 92–113. DOI:10.1080/15267431.2013.768246.

Aldrich, John H., and Forrest D. Nelson. 1984. *Linear Probability, Logit, and Probit Models.* Newbury Park, CA: Sage.

Allen, Jeff, and Huy Le. 2008. "An Additional Measure for Overall Effect Size for Logistic Regression Models." *Journal of Educational and Behavioral Statistics*, 33: 416–441. DOI:10.3102/1076998607306081.

Allison, P. D. 1999. "Comparing Logit and Probit Coefficients Across Groups." *Sociological Methods and Research*, 28: 186–208. DOI:10.1177/0049124199028 002003.

Allison, Paul. 2012. "Logistic Regression for Rare Events." *Statistical Horizons*. Retrieved at http://statisticalhorizons.com/logistic-regression-for-rare-events.

Allison, Paul. 2013. "Why I Don't Trust the Hosmer-Lemeshow Test for Logistic Regression." *Statistical Horizons*. Retrieved at http://statisticalhorizons.com/hosmer-lemeshow.

Audrain-McGovern, Janet, Kenneth P. Tercyak, Alexandra E. Shields, Angelita Bush, Carlos Francisco Espinel, and Caryn Lerman. 2003. "Which Adolescents are Most Receptive to Tobacco Industry Marketing? Implications for Counter-Advertising Campaigns." *Health Communication*, 15: 499–513. DOI:10.1207/S15327027HC1504_07.

Avery, Elizabeth Johnson, and Ruthann Weaver Lariscy. 2014. "Preventable Disease Practices among a Lower SES. Multicultural, Nonurban U.S. Community: The Roles of Vaccination Efficacy and Personal Constraints." *Health Communication*, 29: 826–836. DOI:10.1080/10410236.2013.804486.

Azen, Razia, and Cindy M. Walker. 2011. *Categorical Data Analysis for the Behavioral and Social Sciences.* New York: Routledge.

Berry, Tanya R., John C. Spence, Ronald C. Plotnikoff, and Adrian Bauman. 2011. "Physical Activity Information Seeking and Advertising Recall." *Health Communication*, 26: 246–254. DOI:10.1080/10410236.2010.549810.

Berry, William D. 1993. *Understanding Regression Assumptions.* Newbury Park, CA: Sage.

Beullens, Kathleen, Keith Roe, and Jan Van den Bulck. 2013. "Driving Game Playing as a Predictor of Adolescents' Unlicensed Driving in Flanders." *Journal of Children and Media*, 7: 307–318. DOI:10.1080/17482798.2012.729151.

Beullens, Kathleen, and Jerry Van den Bulck. 2013. "Predicting Young Drivers' Car Crashes: The Role of Music Video Viewing and the Playing of Driving Games. Results from a Prospective Cohort Study." *Media Psychology*, 16: 88–114. DOI:10. 1080/15213269.2012.754582.

Bliss, C. I. 1935. "The Calculation of the Dosage-Mortality Curve." *Annals of Applied Biology*, 22: 134–167. DOI:10.1111/j.1744-7348.1935.tb07713.x.

Box, G. E. P., and D. R. Cox. 1964. "An Analysis of Transformations." *Journal of the Royal Statistical Society, Series B*, 26: 211–252.

Box, G. E. P., and P. W. Tidwell. 1962. "Transformation of the Independent Variables." *Technometrics*, 4: 531–550. DOI:10.1080/00401706.1962.10490038.

Caffo, B., and M. Griswold. 2006. "A User-Friendly Introduction to Link-Probit-Normal Models." *American Statistician*, 60: 139–145. DOI:10.1198/00031300 6X110203.

Chang, Byeng-Hee, Seung-Eun Lee, and Byoung-Sun Kim. 2006. " Exploring Factors Affecting the Adoption and Continuance of Online Games among College Students in South Korea: Integrating Uses and Gratification and Diffusion of Innovation Approaches." *New Media & Society*, 8: 295–319. DOI:10.1177/1461444806059888.

Collins, Rebecca L. 2008. "Media Multitasking: Issues Posed in Measuring the Effects of Television Sexual Content Exposure." *Communication Methods and Measures*, 2: 65–79. DOI:10.1080/19312450802063255.

Cox, D. R. 1970. *The Analysis of Binary Data*. London: Spottiswoods, Ballantyne & Co.

Darst, Burcu F., Lisa Madlensky, Nicholas J. Schork, Eric J. Topol, and Cinnamon S. Bloss. 2014. "Characteristics of Genomic Test Consumers Who Spontaneously Share Results with Their Health Care Provider." *Health Communication*, 29: 105–108. DOI:10.1080/10410236.2012.717216.

Davis, Kevin C., James Nonnemaker, Jennifer Duke, and Matthew C. Farrelly. 2013. "Perceived Effectiveness of Cessation Advertisements: The Importance of Audience Reactions and Practical Implications for Media Campaign Planning." *Health Communication*, 28: 461–472. DOI:10.1080/10410236.2012.696535.

DeMaris, Alfred. 1992. *Logit Modeling: Practical Applications*. Newbury Park, CA: Sage.

Denham, Bryan E. 2012. "NY Times War on Drugs Sources Change after September 11." *Newspaper Research Journal*, 33(4): 34–47.

Denham, Bryan E. 2014. "Intermedia Attribute Agenda-Setting in the New York Times: The Case of Animal Abuse in U.S. Horse Racing." *Journalism & Mass Communication Quarterly*, 91: 17–37. DOI:10.1177/1077699013514415.

Denk, Charles E., and Steven E. Finkel. 1992. "The Aggregate Impact of Explanatory Variables in Logit and Linear Probability Models." *American Journal of Political Science*, 36: 785–804.

Deutsch, Tobias, Thomas Frese, and Hagen Sandholzer. 2014. "Factors Associated with Family-Centered Involvement in Family Practice – A Cross-Sectional Multivariate Analysis." *Health Communication*, 29: 689–697. DOI:10.1080/10410236.2013.773409.

Dixon, Helen, Charles Warne, Maree Scully, Suzanne Dobbinson, and Melanie Wakefield. 2014. "Agenda-Setting Effects of Sun-Related News Coverage on Public Attitudes and Beliefs about Tanning and Skin Cancer." *Health Communication*, 29: 173–181. DOI:10.1080/10410236.2012.732027.

Dobransky, Kerry, and Eszter Hargittai. 2012. "Inquiring Minds Acquiring Wellness: Uses of Online and Offline Sources for Health Information." *Health Communication*, 27: 331–343. DOI:10.1080/10410236.2011.585451.

Dunleavy, V. Orrego, L. Crandall, and L. R. Metsch. 2005. "A Comparative Study of Sources of Health Information and Access to Preventive Care among Low Income Chronic Drug Users." *Communication Research Reports*, 22: 117–128. DOI:10.1080/00036810500130554.

Dunlop, Sally M., Trish Cotter, and Donna Perez. 2014. "When Your Smoking is Not Just about You: Antismoking Advertising, Interpersonal Pressure, and Quitting Outcomes." *Journal of Health Communication*, 19: 41–56. DOI:10.1080/10810730.2013.798375.

Dunlop, Sally M., Melanie Wakefield, and Yoshihisa Kashima. 2008. "The Contribution of Antismoking Advertising to Quitting: Intro- and Interpersonal Processes." *Journal of Health Communication*, 13: 250–266. DOI:10.1080/10810730801985301.

Dylko, Ivan B. 2010. "An Examination of Methodological and Theoretical Problems Arising from the Use of Political Participation Indexes in Political Communication Research." *International Journal of Public Opinion Research*, 22: 523–534. DOI:10.1093/ijpor/edq032.

Fairlie, Anne M., Kristen J. Quinlan, William Dejong, Mark D. Wood, Doreen Lawson, and Caren Francione Witt. 2010. "Sociodemographic, Behavioral, and Cognitive Predictors of Alcohol-Impaired Driving in a Sample of U.S. College Students." *Journal of Health Communication*, 15: 218–232.

Faulkner, Guy E. J., Matthew Y. W. Kwan, Margaret MacNeill, and Michelle Brownrigg. 2011. "The Long Live Kids Campaign: Awareness of Campaign Messages." *Journal of Health Communication*, 16: 519–532. DOI: 10.1080/10810730.2010.546489.

Feeley, Thomas Hugh, Vivian M. Williams, and Timothy J. Wise. 2005. "Testing the Predictive Validity of the GRE Exam on Graduate Student Success: A Case Study at University of Buffalo." *Communication Quarterly*, 53: 229–245. DOI:10.1080/01463370500090209.

Fienberg, Stephen E. 2007. *The Analysis of Cross-Classified Categorical Data*, 2nd edition. New York: Springer.

Hagle, Timothy M., and Glenn E. Mitchell, II. 1992. "Goodness-of-Fit Measures for Probit and Logit." *American Journal of Political Science*, 36: 762–784.

Harada, Kazuhiro, Kei Hirai, Hirokazu Arai, Yoshiki Ishikawa, Jun Fukuyoshi, Chisato Hamashima, Hiroshi Saito, and Daisuke Shibuya. 2013. "Worry and Intention among Japanese Women: Implications for an Audience Segmentation Strategy to Promote Mammography Adoption." *Health Communication*, 28: 709–717. DOI:1 0.1080/10410236.2012.711511.

Haun, Jolie, Stephen Luther, Virginia Dodd, and Patricia Donaldson. 2012. "Measurement Variation Across Health Literacy Assessments: Implications for Assessment Selection in Research and Practice." *Journal of Health Communication*, 17: 141–159. DOI:10.1080/10810730.2012.712615.

Hosmer, David W., and Stanley Lemeshow. 2000. *Applied Logistic Regression*, 2nd edition. New York: John Wiley & Sons.

Huang, Chi, and Todd G. Shields. 2000. "Interpretation of Interaction Effects in Logit and Probit Analyses: Reconsidering the Relationship Between Registration Laws, Education, and Voter Turnout." *American Politics Quarterly*, 28: 80–95. DOI:10. 1177/1532673X00028001005.

Imrey, Peter B., Gary G. Koch, and Maura E. Stokes. 1981. "Categorical Data Analysis: Some Reflections on the Log Linear Model and Logistic Regression. Part I: Historical and Methodological Overview." *International Statistical Review*, 49: 265–283.

Imrey, Peter B., Gary G. Koch, and Maura E. Stokes. 1982. "Categorical Data Analysis: Some Reflections on the Log Linear Model and Logistic Regression. Part II: Data Analysis. *International Statistical Review*, 50: 35–63.

Jaccard, James. 2001. *Interaction Effects in Logistic Regression*. Thousand Oaks, CA: Sage.

Johnston, Lloyd D., Jerald G. Bachman, Patrick M. O'Malley, and John Schulenberg. 2012. Monitoring the Future: A Continuing Study of American Youth. Funded by

National Institute on Drug Abuse. Institutional for Social Research, University of Michigan.

Kenney, Shannon R., Joseph W. LaBrie, and Andrew Lac. 2013. "Injunctive Peer Misperceptions and the Mediation of Self-Approval on Risk for Driving after Drinking among College Students." *Journal of Health Communication*, 18: 459–477. DOI:10.1080/10810730.2012.727963.

Kim, Kyunghye, and Nahyun Kwon. 2010. "Profile of e-Patients: Analysis of Their Cancer Information-Seeking from a National Survey." *Journal of Health Communication*, 15: 712–733. DOI:10.1080/10810730.2010.514031.

Kim, Sei-Hill, and James Shanahan. 2003. "Stigmatizing Smokers: Public Sentiment Toward Cigarette Smoking and its Relationship to Smoking Behaviors." *Journal of Health Communication*, 8: 343–367. DOI:10.1080/10810730305723.

King, Gary, and Langche Zeng. 2001. "Logistic Regression in Rare Events Data." *Political Analysis*, 9: 137–163.

Kleinbaum, D. G. 1992. *Logistic Regression: A Self Learning Text*. New York: Springer.

Kleinnijenhuis, Jan, Anita M. J. van Hoof, Dirk Oegema, and Jan A. de Ridder. 2007. "A Test of Rivaling Approaches to Explain News Effects: News on Issue Positions of Parties, Real-World Developments, Support and Criticism, and Success and Failure." *Journal of Communication*, 57: 366–384. DOI:10.1111/j.1460-2466.2007.00347.x.

Koskan, Alexis M., Kamilah B. Thomas-Purcell, Daohai Yu, Gwendolyn P. Quinn, Sophie Dessureault, David Shibata, Paul B. Jacobsen, and Clement K. Gwede. 2014. "Discussion of First-degree Relatives' Colorectal Cancer Risk: Survivors' Perspectives." *Health Communication*, 29: 782–790. DOI:10.1080/10410236.2013.796871.

Kwak, Hyokjin, Richard J. Fox, and George M. Zinkhan. 2002. "What Products Can Be Successfully Promoted and Sold Via the Internet?" *Journal of Advertising Research*, 42: 23–38.

Lee, Ki-Young, and Sung-Hee Joo. 2005. "The Portrayal of Asian Americans in Mainstream Magazine Ads: An Update." *Journalism & Mass Communication Quarterly*, 82: 654–671. DOI:10.1177/107769900508200311.

Liao, T. F. 1994. *Interpreting Probability Models: Logit, Probit, and Other Generalized Linear Models*. Thousand Oaks, CA: Sage.

Martins, Nicole, and Barbara J. Wilson. 2011. "Parental Communication about Kidnapping Stories in the US News." *Journal of Children and Media*, 5: 132–146. DOI:10.1080/17482798.2011.558261.

Mello, Susan, Andy S. L. Tan, Katrina Armstrong, J. Sanford Schwartz, and Robert C. Hornik. 2013. "Anxiety and Depression among Cancer Survivors: The Role of Engagement with Sources of Emotional Support Information." *Health Communication*, 28: 389–396. DOI:10.1080/10410236.2012.690329.

Menard, Scott. 2002. *Applied Logistic Regression*, 2nd edition. Thousand Oaks, CA: Sage.

Meyer, Janet R., and Kyra Rothenberg. 2004. "Repairing Regretted Messages: Effects of Emotional State, Relationship Type, and Seriousness of Offense." *Communication Research Reports*, 21: 348–356. DOI:10.1080/08824090409359999.

Miller, Claude H., Michael Burgoon, Joseph R. Grandpre, and Eusebio M. Alvaro. 2006. "Identifying Principal Risk Factors for the Initiation of Adolescent Smoking Behaviors: The Significance of Psychological Reactance." *Health Communication*, 19: 241–252. DOI:10.1207/s15327027hc1903_6.

Moorman, Sara M. 2011. "The Importance of Feeling Understood in Marital Conversations about End-of-life Health Care." *Journal of Social and Personal Relationships*, 28: 100–116. DOI:10.1177/0265407510386137.

Morgan, Susan E. 2004. "The Power of Talk: African Americans' Communication with Family Members about Organ Donation and Its Impact on the Willingness to Donate Organs." *Journal of Social and Personal Relationships*, 21: 112–124. DOI:10.1177/0265407504039845.

Napoli, Philip M. 2000. "The Federal Communications Commission and Broadcast Policy-Making – 1966–95: A Logistic Regression Analysis of Interest-Group Influence." *Communication Law and Policy*, 5: 203–233. DOI:10.1207/S15326926CLP0502_3.

Niederdeppe, Jeff, Dominick L. Frosch, and Robert C. Hornik. 2008. "Cancer News Coverage and Information Seeking." *Journal of Health Communication*, 13: 181–199. DOI:10.1080/10810730701854110.

Norusis, Marija J. 2005. *SPSS 14.0 Advanced Statistical Procedures Companion*. Upper Saddle River, NJ: Prentice-Hall.

Pampel, Fred C. 2000. *Logistic Regression: A Primer*. Thousand Oaks, CA: Sage.

Pregibon, D. 1981. "Logistic Regression Diagnostics." *Annals of Statistics*, 9: 705–724.

Robinson, James D., and Yan Tian. 2009. "Cancer Patients and the Provision of Informational Social Support." *Health Communication*, 24: 381–390. DOI:10.1080/10410230903023261.

Simonoff, Jeffrey S. 1998. "Logistic Regression, Categorical Predictors, and Goodness-of-Fit: It Depends on Who You Ask." *The American Statistician*, 53: 10–14. DOI: 10.1080/00031305.1998.10480529.

Stokes, Maura E., Charles S. Davis, and Gary G. Koch. 2012. *Categorical Data Analysis Using SAS*, 3rd edition. Cary, NC: SAS.

Tansey, Richard, Michael White, Rebecca G. Long, and Mark Smith. 1996. "A Comparison of Loglinear Modeling and Logistic Regression in Management Research." *Journal of Management*, 22: 339–358. DOI:10.1177/014920639602200207.

Taylor-Clark, Kalahn Alexandra, Kasisomayajula Viswanath, and Robert J. Blendon. 2010. "Communication Inequalities During Public Health Disasters: Katrina's Wake." *Health Communication*, 25: 221–229. DOI:10.1080/10410231003698895.

The American National Election Studies. 2008. American National Election Study: ANES Pre- and Post-Election Survey. ICPSR25383-v2. Ann Arbor, MI: Inter-university Consortium for Political and Social Research [distributor], 2012-08-30. DOI:10.3886/ICPSR25383.v2.

Theil, Henri. 1970. "On the Estimation of Relationships Involving Qualitative Variables." *American Journal of Sociology*, 76: 103–154.

Witting, Marjon, Magda M. Boere-Boonekamp, Margot A. H. Fleuren, Ralph J. B. Sakkers, and Maarten J. IJzerman. 2012. "Predicting Participation in Ultrasound

Hip Screening from Message Framing." *Health Communication*, 27: 186–193. DOI: 10.1080/10410236.2011.571760.
Young, Dannagal Goldthwaite. 2006. "Late-night Comedy and the Salience of the Candidates' Caricatured Traits in the 2000 Election." *Mass Communication & Society*, 9: 339–366. DOI:10.1207/s15327825mcs0903_5.

7

Multinomial Logistic Regression

Chapter 6 addressed binary logistic regression, a technique used in analyses containing a dichotomous response variable. The current chapter addresses multinomial logistic regression, used when a nominal response measure contains more than two categories (Hosmer and Lemeshow 2000, Liao 1994, Menard 2002; see, earlier, Theil 1969, 1970). As an example of an unordered polytomous variable, a researcher analyzing news photographs of political leaders might classify facial expressions as positive, negative, or neutral, reporting on whether valance varies across news organizations as well as individuals pictured. A survey researcher might ask respondents about a primary news source, with outcome categories including print media, television news, the Internet, radio, or another type of source. In a sports setting, a scholar conducting a field study of communication between coaches and athletes might code utterances as task-specific, non-task-specific, or as communication unrelated to athletic competition, as defined in the study.

As explained in the chapter, in addition to analyses containing nominal outcome measures, the multinomial model often proves useful in studies containing ordinal dependent variables, as some analyses do not meet the assumptions of ordinal logistic regression (discussed in Chapter 8). In discussing regression models containing multi-category response measures, the chapter draws on data gathered in the 2008 American National Election Study and the 2012 Monitoring the Future study. The following section reviews multinomial regression applications in published communication research.

Examples of Published Research

Relatively few scholars have used multinomial logistic regression analyses in communication studies. Of those who have, nearly all have analyzed processes in health communication. Jones, Denham, and Springston (2007), for instance, analyzed the influence of mass and interpersonal communication on whether respondents underestimated, estimated correctly, or overestimated their risk of developing breast cancer. Pilgrim and her colleagues (2014) tested the explanatory power of scales such as Clinician-Client Centeredness and Clinic Discomfort on perceptions of care and satisfaction with services in healthcare settings. Chou et al. (2010) applied multinomial analyses in a study of healthcare satisfaction ratings, while Kelly and her colleagues (2009) used the multinomial model in research involving information seeking among underserved populations. Mathur, Levy, and Royne (2013) used multinomial regression to test the effects of demographic variables and smoking status on trust in doctors and family members as sources of cancer information, while Denham (2014) analyzed the effects of sex, race, student activity levels, and parental communication on alcohol risk perceptions among adolescents. In non-health-related research, Ji, Ha, and Sypher (2014) analyzed demographic variables and media-use measures as predictors of information overload, with respondents indicating whether they frequently experienced a sense of overload, sometimes experienced such a sense, or did not feel overloaded by information at any point. Earlier, in a study of public opinion, Schmitt-Beck (1996) used the multinomial model to examine effects of communication on vote choice. The following section discusses the fundamental components of multinomial logistic regression.

Multinomial Logistic Regression: Fundamentals

The previous chapter discussed the logit, or log of the odds, as a link function in binary logistic regression. Using notation from Norusis (2005, 44), the binary model can be expressed as:

$$\log\left(\frac{P(event)}{P(no\,event)}\right) = \beta_0 + \beta_1 X_1 + \beta_2 X_2 + ... + \beta_p X_p$$

where β_0 represents the intercept, β_1 to β_p represent logistic regression coefficients, and X_1 to X_p refer to independent variables. For practical purposes, one can view multinomial logistic regression as an extension of the binary model, with categories of the response measure analyzed simultaneously in reference to a baseline category (see Agresti 1990, 306–317, DeMaris 2004, 294–302, Long 1997, 148–186).[1] Where the binary model produces one set of parameter estimates, the multinomial model produces separate log odds for each set of relationships analyzed. If, for example, a response measure contains three categories, the multinomial model will calculate log odds and a distinct intercept for category 1 relative to category 3 and for category 2 relative to category 3. Thus, if a dependent variable contains J categories, a multinomial model will contain $J-1$ logits yielding distinct parameter estimates and intercept values.[2] Given those processes, the multinomial logistic regression model is expressed as:

$$\log\left(\frac{P\left(category\,i\right)}{P\left(category\,J\right)}\right) = \beta_{io} + \beta_{i1}X_1 + \beta_{i2}X_2 + \ldots + \beta_{ip}X_p$$

where J serves as the baseline category for the ith response option. In this model the first subscript value identifies the corresponding logit and the second the variable (Norusis 2005, 44).

While the binary model follows a binomial distribution (see Chapter 1), polytomous regression models follow a multinomial distribution, with a model estimating the log odds of an observation appearing in a specific category of the response measure. In the multinomial model, maximum likelihood establishes parameter estimates, and a generalized logit serves as the link function. As with other procedures addressed in this text, the multinomial model assumes response categories are mutually exclusive and exhaustive, with every observation assigned to one – and only one – category of the dependent variable. The following section demonstrates parameter estimation in a model containing one three-category explanatory measure and one three-category response variable.

Simple Multinomial Logistic Regression Analysis

Table 7.1 contains cross-tabulated data gathered in the 2008 American National Election Study (The American National Election Studies 2008). In this table, three categories of race (White, Black, Other Race) appear in the rows and three categories of political party affiliation (Democrat, Republican, Independent) appear in the columns. Row percentages indicate that Black respondents and members of other races identified more frequently as Democrats, just as Whites identified more frequently as both Republicans and Independents. Relatively few Black respondents and members of other races identified as Republicans.

Table 7.2 contains a simple multinomial logistic regression analysis, with race entered as an explanatory measure and political party affiliation entered as a response variable. As shown in the table, separate logits appear for Democrats and Republicans with Independents as the reference, or baseline, category. The third category in the race measure, Other Race, served as a reference category for White and Black respondents. Examining estimates in Democrat categories, the parameter

Table 7.1   Cross-tabulation of race by political party affiliation

| | Political Party Affiliation | | | |
Race	Democrat	Republican	Independent	Totals
White	*a* 193 (30.2%)	*b* 189 (29.6%)	*c* 257 (40.2%)	639
Black	*d* 215 (74.4%)	*e* 4 (1.4%)	*f* 70 (24.2%)	289
Other Race	*g* 67 (54.5%)	*h* 12 (9.8%)	*i* 44 (35.8%)	123
Totals	475	205	371	1,051

Table 7.2   Simple multinomial logistic regression model testing race as a determinant of political party affiliation

Response Categories	Explanatory Categories	B	SE	Wald	df	Sig.	Exp(B)
Democrat	Intercept	.421	.194	4.696	1	.030	
	White	–.707	.216	10.694	1	.001	.493
	Black	.702	.238	8.699	1	.003	2.017
Republican	Intercept	–1.299	.326	15.917	1	.001	
	White	.992	.339	8.538	1	.003	2.696
	Black	–1.563	.609	6.596	1	.010	.210

value of – .707 for White respondents exponentiated to .493, indicating the odds of White respondents identifying themselves as Democrats relative to Independents were about half the same odds for members of another race, apart from Blacks. To observe those odds, one can calculate cross-product ratios based on corresponding cell frequencies in Table 7.1. Specifically, one can multiply counts in cells *a* (193) and *i* (44) and divide by the product of counts in cells *c* (257) and *g* (67). Thus, (193)(44)/(257)(67)=8,492/17,219=.493. Similarly, in observing an odds ratio of 2.017 for Black respondents relative to individuals from other races, one can multiply counts in cells *d* (215) and *i* (44) and divide by the product of counts in cells *f* (70) and *g* (67). Thus, (215)(44)/(70)(67)=9,460/4,690=2.017. Given the estimates appearing in Table 7.2, linear models would be expressed as:

$$\log\left(\frac{P(democrat)}{P(independent)}\right) = .421 - .707 + .702$$

$$\log\left(\frac{P(republican)}{P(independent)}\right) = -1.299 + .992 - 1.563$$

As with binary logistic regression, the multinomial model produces likelihood values that inform researchers of (a) whether a model containing explanatory measures performs better than one containing only the intercept, and (b) whether each explanatory measure makes a significant contribution. In the current example, the –2 log likelihood value of a model containing the intercept only was 396.010, and with the addition of race, the value moved to 177.172. The chi-square difference, 218.838 with 4 degrees of freedom, showed significance at $p<.001$, and thus the model containing race performed better than a model containing only the intercept.

In addition to likelihood values, multinomial logistic regression reports three types of pseudo R-square measures (Cox and Snell, Nagelkerke, McFadden) as well as the Hosmer and Lemeshow goodness-of-fit test. In the present example, a non-significant p-value for the Hosmer and Lemeshow test indicated an acceptable fit to the data, and pseudo R-square values ranged from .100 for McFadden to .214 for Nagelkerke. As indicated in Chapter 6, one limitation of pseudo R-squared measures is that one cannot compare them across samples (Mood 2010). Overall, though, race appears to have made both statistically significant and conceptually substantive contributions to notions of political party affiliation.

Multiple Multinomial Logistic Regression Analysis

Building on the ANES example from the previous section, Table 7.3 contains estimates for a multinomial regression model that includes three categorical measures, sex, race, and military service, as well as one continuous measure,

Table 7.3 Multiple multinomial logistic regression model testing sex, race, military service, and newspaper exposure as determinants of political party affiliation

Response Categories	*Explanatory Categories*	*B*	*SE*	*Wald*	*df*	*Sig.*	*Exp(B)*
Democrat	Intercept	.542	.211	6.582	1	.010	
	Newspaper	.053	.028	3.539	1	.060	1.054
	Male	–.612	.155	15.475	1	.001	.542
	White	–.726	.219	10.975	1	.001	.484
	Black	.709	.241	8.645	1	.003	2.031
	Military	.185	.219	.713	1	.398	1.203
Republican	Intercept	–1.491	.344	18.815	1	.000	
	Newspaper	.092	.032	7.987	1	.005	1.096
	Male	–.111	.190	.344	1	.558	.895
	White	.931	.341	7.444	1	.006	2.537
	Black	–1.609	.610	6.961	1	.008	.200
	Military	.397	.248	2.558	1	.110	1.487

newspaper exposure, as determinants of political party affiliation. In addition to the sex and race measures, the ANES asked respondents to indicate the number of days they had read a newspaper during the previous week, and the military item indicated whether individuals had not served or had served/ were currently serving.

Examining Table 7.3, the continuous explanatory measure, newspaper exposure, appears first in both lists of estimates. An important assumption in logistic regression is that continuous explanatory variables show linearity in the log odds (DeMaris 2004, 287). In this case, one might expect political partisans to read the newspaper more frequently than other individuals, meaning parameter estimates would move in a linear direction for each day of the week. One approach for testing the linearity assumption is to enter a continuous explanatory measure as a categorical variable and examine resulting parameter estimates. If they do not show a specific pattern (e.g., move from lower to higher values), then the variable should be treated as categorical instead of continuous. In the current analysis, newspaper exposure satisfied the linearity assumption and appears in Table 7.3 as a continuous predictor. While it missed showing significance in the first series of estimates, it appeared significant in the second, with newspaper reading increasing the odds of individuals identifying themselves as Republicans.

Looking to the categorical determinants in Table 7.3, the odds of males identifying as Democrats relative to Independents were .542 times the same odds for females. Race measures changed very little, with Whites significantly less likely to identify as Democrats and Black respondents significantly more likely to do so. Military service did not prove significant as a determinant. In the second logit analysis, differences did not emerge across males and females nor did military service show significance. Race again proved significant, though, with the odds of White respondents identifying as Republicans relative to Independents 2.537 times the same odds for members of other races. The odds of Blacks identifying as Republicans relative to Independents were .200 times the same odds for members of other races.

Regarding fit, a multinomial model containing the intercept only showed a –2 log likelihood score of 629.293, with the final model reducing that score to 381.160. The chi-squared difference, 248.133 with 10 degrees of freedom, showed significance at $p < .001$. Individual likelihood tests showed significance for sex, race, and newspaper exposure; military service did not prove significant and added little to the multinomial model. Pseudo R-square measures ranged from .113 for McFadden to .240 for Nagelkerke, and in examining model statistics as a whole, race clearly played the most important role.

Conditional Logit Modeling

Closely related to multinomial logistic regression is the conditional logit, or discrete-choice, model. Developed by McFadden (1973), conditional logit analysis considers as explanatory measures the characteristics of choice options as opposed to (or in addition to) the characteristics of individuals making a choice (see Glasgow 2004, Hoffman and Duncan 1988, 415). "In the standard multinomial model," Powers and Xie (2000, 239) explained, "explanatory variables are invariant within outcome categories, but their parameters vary with outcome. In the conditional logit model, explanatory variables vary by outcome as well as by the individual, whereas their parameters are assumed constant over all outcome categories."

In public opinion research, the conditional logit model has proven useful in analyses of panel data, which often contain more than one response from the same individual. As an example, Yanovitzky and Capella (2001) studied the effects of political talk radio on listener attitudes during the 1996 US presidential election. The conditional logit model helped to determine whether (a) radio programming affected attitudes or (b) individuals with existing political biases chose certain programs. The authors pooled multiple responses from individuals into one sample, resulting in a series of "person-time" units for each wave in the study. As Yanovitzky and Capella (2001, 386) explained, "Since information on fixed characteristics such as age, gender, race, and political affiliation is duplicated for each individual in each wave, all fixed characteristics (observed and unobserved) that may account for attitude change across waves are controlled for and the model can only be estimated from time-dependent variables..." The conditional logit model, the authors noted, eliminated concerns about spuriousness of causal relationships between exposure to political talk radio and attitude change.

In political science, Baumgartner (2012) used the conditional logit model to forecast the selection of vice-presidential candidates, 1960 to 2008. In each election cycle, multiple candidates emerged in both major parties, with one Democrat and one Republican ultimately selected. Baumgartner positioned "nominated" as a dependent variable, using conditional logit analysis to identify media exposure, political experience, having served in the military, age, and gender/racial/ethnic diversity as factors associated with being nominated. The characteristics of choices, rather than individuals making the selections, proved central to the analysis.

In communication, the conditional logit model may prove beneficial to scholars who work with panel data gathered during political and health campaigns. Dating back to the election research of Lazarsfeld and his colleagues, panel data has been used in analyses involving media use as well as interpersonal communication.

By considering differences in the values assigned to choice options across time, the technique can detect changes in attitudes following substantive events.[3] To date, the procedure has been used in political science more frequently than in communication, where no studies apart from public opinion analyses have emerged.

Multinomial Logistic Regression in SPSS

This section of the chapter uses data gathered in the 2012 Monitoring the Future study (Johnston et al. 2012) to demonstrate multinomial logistic regression analysis in SPSS. A three-category dependent variable indicated whether MTF respondents did not disapprove, disapproved, or strongly disapproved of individuals 18 or over trying the illicit substance MDMA (i.e., "ecstasy"). MDMA is considered a "club drug" and is often used by individuals attending raves, or all-night dance parties. Three categorical factors served as determinants of attitudes toward the substance: sex of respondents, the perceived ease with which MDMA could be obtained, and exposure to anti-drug spots on television. Respondents indicated whether it was probably impossible, very difficult, fairly difficult, fairly easy, or very easy to obtain MDMA, also indicating whether they had not seen anti-drug spots or had seen them less than once per month, one to three times per month, one to three times per week, or daily.

To begin a multinomial analysis in SPSS, a researcher should first click on Analyze, followed by Regression and Multinomial. At that point SPSS allows the analyst to enter explanatory and response variables, as shown in Figure 7.1. For the response measure, a researcher can select a baseline category for logit analyses; the last category serves as the default, but one can also choose the first category or customize the analysis. Categorical explanatory variables should be entered as Factors and continuous determinants as Covariates. In general, ordinal measures should be entered first as Factors, as one cannot assume linearity in the respective logits and equal intervals between observation points.[4] When linearity is satisfied and a measure contains at least four categories, then it may be tested as a Covariate.

Figure 7.1 also contains a window for model construction. Here one can choose from Main effects, Full factorial, and Custom/Stepwise. Main effects models contain no interactions, while a Full factorial model contains all potential interactions among explanatory variables. In many cases, a researcher may seek to customize a multinomial analysis by entering one interactive term instead of every potential interaction, or examine model estimates with variables added at different points; in such cases the Custom/Stepwise option may prove useful.

Figure 7.2 contains output options for multinomial logistic regression. As the window on the right indicates, the procedure includes multiple options, and

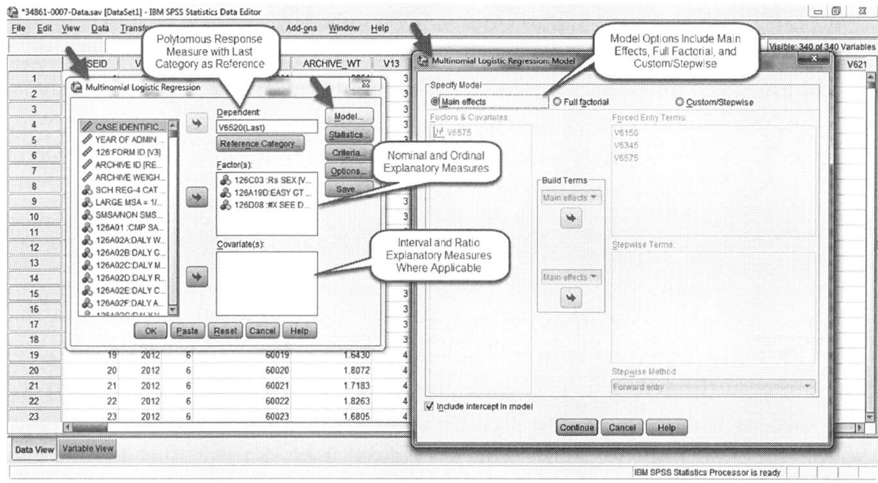

Figure 7.1   SPSS screenshots for variables to be included in multinomial logistic regression model. Source: SPSS® Reprints Courtesy of International Business Machines Corporation, © 2014 International Business Machines Corporation

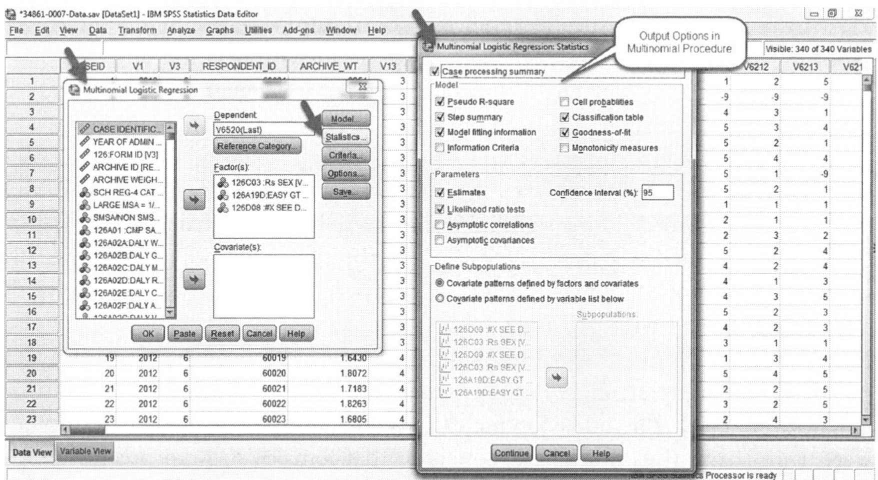

Figure 7.2   SPSS screenshots for output options in multinomial logistic regression model. Source: SPSS® Reprints Courtesy of International Business Machines Corporation, © 2014 International Business Machines Corporation

especially important are parameter estimates, likelihood ratio tests, goodness-of-fit tests, pseudo R-square measures, and the 95% confidence interval. A researcher should select and review these statistics with each analysis. Step summaries can also prove useful for identifying the relative importance of explanatory factors.

Although the current chapter does not include separate screenshots for the Criteria, Options, and Save windows – default settings generally perform well and do not need adjustment – it should be noted that in a multivariate analysis, the number of cells can grow rapidly, introducing potential problems with empty cells. In the Options window, a researcher should set the value of Delta at .5 when zero-count cells emerge. SPSS will then add .5 to each cell, ideally adding stability to parameter estimates. Empty cells are especially common in analyses with small samples.

Table 7.4 contains SPSS output for the multinomial model testing sex, ease of obtaining MDMA, and exposure to anti-drug spots as predictors of attitudes toward MDMA experimentation. The first set of statistics compares the model containing all three explanatory measures to a model containing the intercept only, and as indicated by the significance of chi-square, the three-variable model offers an improvement. The goodness-of-fit statistics just below that information confirm the fit, with a nonsignificant chi-square value showing no differences between observed frequencies and those produced by the main-effects multinomial model.

Continuing with Table 7.4, pseudo R-square estimates range from .062 to .128 and, examining the results of likelihood ratio tests, sex and ease of obtaining MDMA both made statistically significant contributions to the model; however, exposure to drug spots did not make a significant contribution and could be dropped in a subsequent analysis. Examining –2 log likelihood statistics and chi-square values, it appears ease of obtaining MDMA made an especially strong contribution, as the model would have shown a –2 log likelihood value of 605.898 without its inclusion. Without sex included, the –2 log likelihood score would have equaled 429.062, also a significant departure from the final model.

Table 7.5 contains parameter estimates for two levels of the dependent variable – Do Not Disapprove of MDMA experimentation and Disapprove of experimentation – with Strongly Disapprove serving as a reference category. Examining the estimates for those who did not disapprove, the odds of males not disapproving to strongly disapproving were 1.789 times the same odds for females. Similarly, the odds for males disapproving to strongly disapproving were 1.455, and thus one might conclude that females showed greater disapproval of MDMA experimentation. Perceived accessibility of MDMA appears after sex in the two lists of parameter estimates, and as indicated by the log odds among those who did not disapprove, perceived ease/difficulty in obtaining MDMA served as a significant determinant of attitudes toward the substance. The odds of those who considered MDMA probably impossible to obtain not disapproving of experimentation were just .079 times the same odds for those who considered MDMA easy to obtain. Similar odds emerged for those who considered MDMA very difficult to obtain. Among those who disapproved of

Table 7.4   SPSS output for multinomial logistic regression model containing categorical predictors

Model Fitting Information

Model	Model Fitting Criteria	Likelihood Ratio Tests		
Model	*–2 Log Likelihood*	*Chi-Square*	*df*	*Sig.*
Intercept Only	643.825			
Final	404.928	238.896	18	.000

Goodness-of-Fit

	Chi-Square	*df*	*Sig.*
Pearson	80.215	80	.472
Deviance	84.988	80	.330

Pseudo R-Square

Cox and Snell	.108
Nagelkerke	.128
McFadden	.062

Likelihood Ratio Tests

Model	Model Fitting Criteria	Likelihood Ratio Tests		
Model	*–2 Log Likelihood*	*Chi-Square*	*df*	*Sig.*
Intercept	404.928[a]	.000	0	
Sex	429.062	24.134	2	.000
Ease of Getting MDMA	605.898	200.970	8	.000
Exposure to Drug Spots	413.305	8.377	8	.397

The chi-square statistic is the difference in –2 log likelihood between the final model and a reduced model.

The reduced model is formed by omitting an effect from the final model. The null hypothesis is that all parameters of that effect are 0.

[a]This reduced model is equivalent to the final model because omitting the effect does not increase the degrees of freedom.

Table 7.5   SPSS output for multinomial logistic regression model containing categorical predictors

Parameter Estimates

	B	SE	Wald	df	Sig.	Exp(B)	95% Confidence Interval for Exp(B) Lower	Upper
DNT DISP.								
Intercept	−1.155	.311	13.788	1	.000			
Males	.582	.142	16.863	1	.000	1.789	1.355	2.362
MDMA Impossible	−2.539	.297	73.109	1	.000	.079	.044	.141
MDMA Very Difficult	−2.466	.271	82.736	1	.000	.085	.050	.145
MDMA Fairly Difficult	−1.563	.208	56.428	1	.000	.210	.139	.315
MDMA Fairly Easy	−.580	.190	9.313	1	.002	.560	.386	.813
Never See Drug Spots	.645	.308	4.370	1	.037	1.906	1.041	3.489
<ONCE/MO	.559	.315	3.145	1	.076	1.749	.943	3.244
1–3/MO	.691	.310	4.957	1	.026	1.995	1.086	3.665
1–3/WK	.553	.337	2.689	1	.101	1.738	.898	3.366
DISAPPRV.								
Intercept	−1.145	.249	21.208	1	.000			
Males	.375	.105	12.801	1	.000	1.455	1.185	1.786
MDMA Impossible	−.802	.219	13.358	1	.000	.448	.292	.689
MDMA Very Difficult	−.350	.200	3.049	1	.081	.705	.476	1.044
MDMA Fairly Difficult	−.087	.193	.206	1	.650	.916	.628	1.337
MDMA Fairly Easy	.246	.196	1.570	1	.210	1.279	.870	1.880
Never See Drug Spots	.238	.219	1.181	1	.277	1.268	.826	1.947
<ONCE/MO	.279	.219	1.621	1	.203	1.322	.860	2.031
1–3/MO	.262	.217	1.455	1	.228	1.300	.849	1.990
1–3/WK	.413	.213	3.185	1	.074	1.511	.960	2.378

Classification

	Predicted			
Observed	*DNT DISP.*	*DISAPPRV.*	*STRG. DIS.*	*Percentage Correct*
DNT DISP.	30	0	250	10.7%
DISAPPRV.	16	0	523	.0%
STRG. DIS.	34	0	1245	97.3%
Overall percent	3.8%	.0%	96.2%	60.8%

experimentation, odds were not as pronounced, but they did move in a consistent pattern, supporting the notion of perceived access as an important determinant of attitudes toward use.

Examining parameter estimates for exposure to anti-drug spots, significant relationships emerged at the Do Not Disapprove response level. Here, the odds of an individual who saw no anti-drug spots not disapproving of MDMA use (compared to strongly disapproving) were 1.906 times the same odds for those who saw anti-drug spots on a daily basis. Log odds for those who saw anti-drug spots less than once per month approached significance, and the odds of those who saw between one and three spots per month not disapproving of MDMA use were 1.995 the same odds for those who saw anti-drug spots on a daily basis. No relationships showed significance at the Disapprove level of the response measure, and thus a researcher would conclude that while anti-drug spot exposure showed potential as an explanatory measure, especially at the Do Not Disapprove level, relationships proved inconsistent.

Lastly, the classification table located beneath the section on parameter estimates revealed that, overall, the model resulted in 60.8% of cases correctly classified. This figure is somewhat modest, but just three variables appeared in the equation, and two of those measures showed consistency as explanatory variables. With additional measures included, predicted percentages would have increased.

Chapter Summary

This chapter addressed multinomial logistic regression analysis, a technique used when a nominal response measure contains more than two categories. The procedure can be viewed as an extension of the binary logistic regression model, with categories of the response measure analyzed simultaneously in reference to a baseline category. In addition to the multinomial model, the chapter also summarized the conditional logit model, which uses characteristics of choice options as explanatory measures. The conditional logit model has proven useful in panel studies, with scholarship appearing in political science and public opinion journals.

Chapter Exercises

1. Use multinomial logistic regression to test the effects of sex and consumption of energy drinks (No Consumption, Less than One energy drink per day, One or More energy drinks per day) on adolescent use of amphetamines (No Use, 1–2 Times, 3–9 Times, 10 or More Times). In your analysis, report –2 log likelihood values for the Intercept Only and Final models, as well as the

Pearson goodness-of-fit test, three R-squared estimates, and the likelihood ratio tests. Also report parameter estimates and their standard errors, in addition to Wald values, significance tests, exponentiated parameter estimates, and 95% confidence intervals. The data for this exercise, drawn from the 2012 Monitoring the Future study (Johnston et al. 2012), appear below.

Energy Drinks per Day	Amphetamine Use	Frequency
Males		
None	0	596
None	1–2 Times	28
None	3–9 Times	27
None	10 or More Times	21
Less than One	0	219
Less than One	1–2 Times	16
Less than One	3–9 Times	20
Less than One	10 or More Times	18
One or More	0	87
One or More	1–2 Times	4
One or More	3–9 Times	6
One or More	10 or More Times	7
Females		
None	0	721
None	1–2 Times	33
None	3–9 Times	25
None	10 or More Times	22
Less than One	0	129
Less than One	1–2 Times	14
Less than One	3–9 Times	11
Less than One	10 or More Times	7
One or More	0	71
One or More	1–2 Times	2
One or More	3–9 Times	5
One or More	10 or More Times	5

2. Use multinomial logistic regression to test the effects of sex, race (White, Black, Other Race), and exposure to the 2008 election campaign in the newspaper (Exposed, Not Exposed) on political party identification (Democrat, Republican, Independent). In your analysis, report −2 log likelihood values for the Intercept Only and Final models, as well as the Pearson goodness-of-fit test, three R-squared estimates, and the likelihood ratio tests. Also report parameter estimates and their standard errors, in addition to Wald values, significance tests, exponentiated parameter estimates, and

95% confidence intervals. The data for this exercise, drawn from the 2008 American National Election Study (The American National Election Studies 2008), appear below.

Race	Exposure to Campaign in Newspaper	Political Party ID	Frequency
Males			
White	Yes	Democrat	35
White	Yes	Republican	59
White	Yes	Independent	60
White	No	Democrat	9
White	No	Republican	10
White	No	Independent	29
Black	Yes	Democrat	8
Black	Yes	Republican	1
Black	Yes	Independent	9
Black	No	Democrat	20
Black	No	Republican	1
Black	No	Independent	12
Other	Yes	Democrat	8
Other	Yes	Republican	2
Other	Yes	Independent	7
Other	No	Democrat	6
Other	No	Republican	1
Other	No	Independent	5
Females			
White	Yes	Democrat	62
White	Yes	Republican	49
White	Yes	Independent	52
White	No	Democrat	29
White	No	Republican	25
White	No	Independent	36
Black	Yes	Democrat	71
Black	Yes	Republican	1
Black	Yes	Independent	13
Black	No	Democrat	21
Black	No	Republican	0
Black	No	Independent	14
Other	Yes	Democrat	13
Other	Yes	Republican	5
Other	Yes	Independent	9
Other	No	Democrat	13
Other	No	Republican	2
Other	No	Independent	5

Notes

1 Generally, statistical software programs establish the last variable category as a baseline, but most programs allow the researcher to switch a reference category to the initial option (see Norusis 2005, Stokes, Davis, and Koch 2012). Researchers can also recode data, scoring the preferred reference category as the highest value (i.e., category label).

2 Although multinomial logistic regression can use a probit link, calculations often prove complicated, and statisticians generally encourage use of the logit in most circumstances (for discussion, see Aldrich and Nelson 1984, 73, Borooah 2002, Campbell and Donner 1989, Dow and Endersby 2004). The probit is a viable option for binary and ordinal logistic regression modeling.

3 SPSS does not contain a menu option for conditional logit analysis, but researchers have tested such models through the Cox regression procedure. For details on how to structure the data, see Garson (2014). For texts that move beyond panel studies in addressing longitudinal analyses of categorical data, see Liang and Zeger (1986), Sutradhar (2014), and Von Eye and Niedermeier (1999).

4 While communication scholars and other social scientists often treat ordered explanatory and response measures as quasi-interval, a benefit of the multinomial procedure is that it allows one to examine parameter estimates as each level of a predictor. Depending on the research area, such detail can prove important, offering greater information than a single coefficient for a continuous covariate can provide.

References

Agresti, Alan. 1990. *Categorical Data Analysis*. New York: John Wiley & Sons.

Aldrich, John H., and Forrest D. Nelson. 1984. *Linear Probability, Logit, and Probit Models*. Newbury Park, CA: Sage.

Baumgartner, Jody C. 2012. "The Post-Palin Calculus: The 2012 Republican Veepstakes." *PS: Political Science and Politics*, 45: 605–609. DOI:10.1017/S1049096512000996.

Borooah, Vani K. 2002. *Logit and Probit: Ordered and Multinomial Models*. Thousand Oaks, CA: Sage.

Campbell, M. K., and A. Donner. 1989. "Classification Efficiency of Multinomial Logistic Regression Relative to Ordinal Logistic Regression." *Journal of the American Statistical Association*, 84: 587–591. DOI:10.1080/01621459.1989.10478807.

Chou, Wen-Ying Sylvia, Lin Chun Wang, Lila J. Finney Rutten, Richard P. Moser, and Bradford W. Hesse. 2010. "Factors Associated with Americans' Ratings of Health Care Quality: What Do They Tell Us about the Raters and the Health Care System?" *Journal of Health Communication*, 15: 147–156. DOI:10.1080/10810730.2010.522692.

DeMaris, Alfred. 2004. *Regression with Social Data: Modeling Continuous and Limited Response Variables*. Hoboken, NJ: John Wiley & Sons.

Denham, Bryan E. (2014). "Adolescent Perceptions of Alcohol Risk: Variation by Sex, Race, Student Activity Levels and Parental Communication." *Journal of Ethnicity in Substance Abuse*, 13: 385–404. DOI:10.1080/15332640.2014.958638.

Dow, Jay K., and James W. Endersby. 2004. "Multinomial Probit and Multinomial Logit: A Comparison of Choice Models for Voting Research." *Electoral Studies*, 23: 107–122. DOI:10.1016/S0261-3794(03)00040-4.

Garson, G. David. 2014. *Logistic Regression: Binomial and Multinomial.* Asheboro, NC: Statistical Associates Publishers.

Glasgow, Garrett. 2004. "Conditional Logit Model." In Michael S. Lewis-Black, Alan Bryman, and Tim Futing Liao (Eds), *The Sage Encyclopedia of Social Science Research Methods* (pp. 166–167). Thousand Oaks, CA: Sage.

Hoffman, Saul D., and Greg J. Duncan. 1988. "Multinomial and Conditional Logit Discrete-Choice Models in Demography." *Demography*, 25: 415–427.

Hosmer, David W., and Stanley Lemeshow. 2000. *Applied Logistic Regression*, 2nd ed. New York: John Wiley & Sons.

Ji, Qihao, Louisa Ha, and Ulla Sypher. 2014. "The Role of News Media Use and Demographic Characteristics in the Prediction of Information Overload." *International Journal of Communication*, 8: 699–714.

Johnston, Lloyd D., Jerald G. Bachman, Patrick M. O'Malley, and John Schulenberg. 2012. Monitoring the Future: A Continuing Study of American Youth. Funded by National Institute on Drug Abuse. Institutional for Social Research, University of Michigan.

Jones, Karyn Ogata, Bryan E. Denham, and Jeffrey K. Springston. 2007. "Differing Effects of Mass and Interpersonal Communication on Breast Cancer Risk Estimated: An Exploratory Study of College Students and Their Mothers." *Health Communication*, 21: 165–175. DOI:10.1080/10410230701307253.

Kelly, Kimberly M., Amy C. Sturm, Kathleen Kemp, Jacquelin Holland, and Amy K. Ferketich. 2009. "How Can We Reach Them? Information Seeking and Preferences for a Cancer Gamily History Campaign in Underserved Communities." *Journal of Health Communication*, 14: 573–589. DOI:10.1080/10810730903089580.

Liang, Kung-Yee, and Scott L. Zeger. 1986. "Longitudinal Data Analysis Using Generalized Linear Models." *Biometrika*, 73: 13–22. DOI:10.1093/biomet/73.1.13.

Liao, Tim Futing. 1994. *Interpreting Probability Models: Logit, Probit, and Other Generalized Linear Models.* Thousand Oaks, CA: Sage.

Long, J. Scott. 1997. *Regression Models for Categorical and Limited Dependent Variables.* Thousand Oaks, CA: Sage.

Mathur, Sunil, Marian Levy, and Marla B. Royne. 2013. "The Role of Cancer Information Seeking Behavior in Developing and Disseminating Effective Smoking Strategies: A Comparison of Current Smokers, Former Smokers, and Never Smokers." *Journal of Communication in Healthcare*, 6: 61–70. DOI:10.1179/175380761 2Y.0000000022.

McFadden, D. 1973. "Conditional Logit Analysis of Qualitative Choice Behavior." In P. Zarembka (Ed.), *Structural Analysis of Discrete Data with Econometric Implications* (pp. 105–135). Cambridge: MIT Press.

Menard, Scott. 2002. *Applied Logistic Regression Analysis,* 2nd ed. Thousand Oaks, CA: Sage.

Mood, Carina. 2010. "Logistic Regression: Why We Cannot Do What We Think We Can Do, and What We Can Do About It." *European Sociological Review*, 26: 67–82. DOI:10.1093/esr/jcp006.

Norusis, Marija J. 2005. *SPSS 14.0 Advanced Statistical Procedures Companion.* Upper Saddle River, NJ: Prentice Hall.

Pilgrim, Nanlesta A., Kathleen M. Cardona, Evette Pinder, and Freya L. Sonenstein. 2014. "Clients' Perceptions of Service Quality and Satisfaction at Their Initial Title X Family Planning Visit." *Health Communication*, 29: 505–515. DOI:10.1080/10410236.2013.777328.

Powers, Daniel A., and Yu Xie. 2000. *Statistical Methods for Categorical Data Analysis*. San Diego, CA: Academic Press.

Schmitt-Beck, Rudiger.1996. "Mass Media, the Electorate, and the Bandwagon Effect: A Study of Communication Effects on Vote Choice in Germany." *International Journal of Public Opinion Research*, 8: 266–291. DOI:10.1093/ijpor/8.3.266.

Stokes, Maura E., Charles S. Davis, and Gary G. Koch. 2012. *Categorical Data Analysis Using SAS*, 3rd ed. Cary, NC: SAS Institute.

Sutradhar, Brajendra. 2014. *Longitudinal Categorical Data Analysis*. New York: Springer.

The American National Election Studies. 2008. American National Election Study: ANES Pre- and Post-Election Survey. ICPSR25383-v2. Ann Arbor, MI: Inter-university Consortium for Political and Social Research [distributor], 2012–08–30. DOI:10.3886/ICPSR25383.v2.

Theil, H. 1969. "A Multinomial Model Extension of the Linear Logit Model." *International Economic Review*, 10: 251–259.

Theil, H. 1970. "On the Estimation of Relationships Involving Qualitative Variables." *American Journal of Sociology*, 76: 103–154.

Von Eye, Alexander, and Keith E. Niedermeier. 1999. *Statistical Analysis of Longitudinal Categorical Data in the Social and Behavioral Sciences*. Mahwah, NJ: Erlbaum.

Yanovitzky, Itzhak, and Joseph N. Capella. 2001. "Effect of Call-In Political Talk Radio Shows on Their Audiences: Evidence from a Multi-Wave Panel Analysis." *International Journal of Public Opinion Research*, 13: 384–397. DOI:10.1093/ijpor/13.4.377.

8

Ordinal Logistic Regression

Up to this point the discussion of logistic regression analysis has focused on binary and multinomial techniques, used in equations with dichotomous and polytomous response variables, respectively. The current chapter focuses on ordinal logistic regression, used when a dependent measure contains ordered categories (Agresti 1984, Anderson 1984, McCullagh 1980). In communication research, ordinal measures include items such as Likert attitude statements, semantic differential scales, grouped behavior frequencies, and self-reported estimates of media use. Likert statements are especially popular and measure attitudes by asking respondents whether they Strongly Agree, Agree, Neither Agree Nor Disagree, Disagree, or Strongly Disagree with specific assertions. Researchers typically assign scores 1 through 5 to the sequential response options, termed "vague quantifiers" (Schaeffer 1991, Griffin 2013). In some cases, researchers treat the scores as discrete, applying ordinal statistics, but in other instances scholars consider the values "quasi-interval," calculating means and standard deviations (see, for discussion, Hayes 2005, Hewes 1978, Norman 2010, Wilson 1971). The present chapter covers the advantages and disadvantages of both approaches.

In addition to individual survey items such as Likert statements, behavioral models also may contain ordered levels. O'Connell (2006) mentioned the Stages of Change model (Prochaska and DiClemente 1983), which includes the categories *precontemplation*, *contemplation*, *preparation*, *action*, and *maintenance*. In carrying out a study grounded in this model, a researcher would need to operationalize each conceptual stage, such that the technique addressed in this chapter, ordinal logistic regression, might be used to analyze relationships. The following section reviews communication studies that have used ordinal logistic regression. Following that review, the chapter addresses model fundamentals and offers instruction on the Polytomous Universal Model (PLUM) in SPSS.

Categorical Statistics for Communication Research, First Edition. Bryan E. Denham.
© 2017 John Wiley & Sons, Inc. Published 2017 by John Wiley & Sons, Inc.
Companion Website: www.wiley.com/go/denham/categorical_statistics

Examples of Published Research

While research in health communication has accounted for most of the disciplinary scholarship reviewed in this text, scholars have applied ordinal logistic regression to studies involving new media, family communication, and sourcing dynamics in science communication, among other subjects. As an example, Chyi and colleagues (2010) used ordinal regression in examining Internet news site satisfaction, with an ordered dependent variable moving from Mostly Dissatisfied to Mostly Satisfied. Relatedly, Chyi and Yang (2009) drew on data gathered by the Pew Research Center in analyzing online news use, which the researchers operationalized along seven ordered levels. In the context of family communication, Evans and his colleagues (2012) examined effects of media messages on parent–child communication about sex, while Strom and Boster (2011) analyzed family support as a determinant of educational attainment. Studying the extent to which scientists had communicated with media personnel, Dunwoody, Brossard, and Dudo (2009) created a three-level ordinal variable consisting of No Contact, Modest Contact, and Frequent Contact. Stempel, Hargrove, and Stempel (2007) used ordinal regression in studying media use and beliefs in 9/11 conspiracy theories. The authors offered respondents three such theories, followed with the response options Very Likely, Somewhat Likely, and Not Likely.

In health-related research, Denham (2010) used ordinal logistic regression in a study measuring risk perceptions of anabolic steroid use. Response options included Little Risk, Moderate Risk, and Great Risk, with explanatory measures including sex, race, newspaper exposure, anti-drug spot exposure, athletic participation, estimated ease of obtaining steroids, perceived peer use of steroids, and estimated use of the drugs by professional athletes. Han, Moser, and Klein (2006) studied the effects of perceived ambiguity on ordinal measures of cancer preventability, risk and worry, and studying comparative risk for colon cancer, Hay, Coups, and Ford (2006) found higher risk estimates among younger respondents as well as those in poorer health and those with a history of cancer in the family, among other factors.

Ordinal Logistic Regression: Fundamentals

Binary logistic regression models estimate the probability of a "success" (i.e., the probability of an observation taking on the higher of two values). In a content analysis, a success may refer to a news report mentioning a specific issue attribute, while in survey research, a success might consist of a participant discussing politics with a family member in the previous 24 hours. Just two possibilities exist when a dependent variable is dichotomous, and a success is therefore easy to define.

But as O'Connell (2006) explained, the term "success" can take on multiple meanings in the presence of an ordinal response measure. For example, in one study a researcher might be interested in the overall strength of an ordinal association, and in another project the focus may shift to specific levels of ordered measures. Accordingly, statisticians have developed a series of logistic regression models for ordinal data, some of which include cumulative (or proportional) odds, adjacent categories, and continuation ratio (see, for discussion, Ananth and Kleinbaum 1997, Agresti 1984, 1989, Clogg and Shihadeh 1994, Holtbrugge and Schumacher 1991, Hosmer and Lemeshow 2000, Sobel 1997, Winship and Mare 1984).

The current chapter focuses on the most popular ordinal regression model, cumulative odds (McCullagh 1980), because it works well with the kinds of questions communication scholars ask, and because SPSS fits this model in its PLUM procedure (Norusis 2005). In SPSS a cumulative odds model indicates the odds of an observation appearing *at or below* a specific category of an ordinal dependent variable; the name of the model – cumulative odds – reflects the fact that odds accumulate across the categories of an ordered response measure.[1] With cases "sequentially partitioned into dichotomous groups" (O'Connell 2006, 28), the cumulative odds model can be viewed as a series of binary logistic regression analyses (see also DeMaris 2004, 303–307, Stokes, Davis, and Koch 2012, 260–274).

Discussing cumulative odds, O'Connell (2006, 31) used the following notation to reflect the probability (π) of an observation appearing at or below the jth category of K possible outcomes ($j = 1, 2,...K - 1$) of an ordinal response measure (Y) given a set of p covariates: $\pi \, (Y \leq j \mid X_1, X_2,...X_p) = \pi_j \, (\underline{x})$. Cumulative probabilities, expressed as (cumulative) logits in the following model, can be exponentiated to observe the odds of an observation appearing at or below a specific category of the response variable:

$$\ln\left(Y_j'\right) = \ln\left(\frac{\pi_j\left(\underline{x}\right)}{1 - \pi_j\left(\underline{x}\right)}\right) = \beta_j + \left(\beta_1 X_1 + \beta_2 X_2 + ...\beta_p X_p\right)$$

where Y represents the odds in a transformed dependent measure. To help clarify this model, the chapter draws on data gathered in the 2012 Monitoring the Future study (Johnston et al. 2012). Table 8.1 contains a cross-tabulation of sex by perceived risk of marijuana use. Examining row percentages accompanying cell frequencies, it appears females tended to perceive greater risk. Approximately 18% of males estimated no risk, compared to 9.2% of females, and while 34.9% of males estimated great risk, nearly 50% of females did so.

Drawing structurally from O'Connell (2006, 32), Table 8.2 extends the frequency data to include proportions, cumulative proportions, cumulative odds,

Table 8.1 Cross-tabulation of sex by marijuana risk perceptions

Sex	None	Slight	Moderate	Great	Totals
	Perceived Risk Associated with Regular Marijuana Use				
Male	*a* 199 (18.1%)	*b* 256 (23.2%)	*c* 262 (23.8%)	*d* 385 (34.9%)	1,102
Female	*e* 99 (9.2%)	*f* 170 (15.8%)	*g* 276 (25.6%)	*h* 532 (49.4%)	1,077
Totals	298	426	538	917	2,179

Table 8.2 Observed data cross-classification of sex by four levels of risk associated with regular marijuana use: frequency (f), proportion (p), cumulative proportion (cp), cumulative odds[a] (co), and Odds Ratios (OR)

Category	None	Slight	Moderate	Great	Totals (f)
	Perceived Risk Associated with Regular Marijuana Use				
Males					
f	199	256	262	385	1,102
p	.1806	.2323	.2377	.3494	1.000
cp	.1806	.4129	.6506	1.000	—
co	.2204	.7033	1.862	—	—
Females					
f	99	170	276	532	1,077
p	.0919	.1578	.2563	.4940	1.000
cp	.0919	.2498	.5060	1.000	—
co	.1012	.3330	1.024	—	—
OR	**2.1779**	**2.112**	**1.8184**	—	—
Totals (f)	298	426	538	917	2,179
cp_{total}	.1368	.3323	.5792	1.000	—

[a] Cumulative odds = Odds($Y_i \leq$ category j)

and odds ratios for sex across perceptions of risk associated with regular marijuana use. In this table, 199 of 1,102 males estimated no risk, resulting in a proportion of .1806, a cumulative proportion of .1806, and cumulative odds of .2204 (.1806/1 − .1806). Next, 256 of 1,102 males indicated slight risk, resulting in a proportion of .2323, a cumulative proportion of .4129, and cumulative odds of .7033 (.4129/1 − .4129). At the third level of the ordinal measure, 262 of 1,102 males estimated moderate risk, resulting in a proportion of .2377, a cumulative proportion of .6506, and cumulative odds of 1.862 (.6506/1 − .6506). The cumulative proportion for the last category equals 1.0 (as it always does), reflecting $K - 1$ cumulative odds calculations.

With calculations for males complete, the next task is to determine the respective odds for females, as shown in Table 8.2, and then divide the cumulative odds for males by those for females at each level of the ordinal response measure. The latter set of calculations results in odds ratios. As an example, looking to the first category, which contains responses for no risk, .2204/.1012 = 2.1779. At the level of slight risk, .7033/.3330 = 2.112, and at the level of moderate risk, 1.862/1.024 = 1.8184. Given those calculations, one is able to conclude that, on average, the odds of a male appearing at or below a given level of the dependent variable are about twice those same odds for females. In other words, males tended to estimate less risk associated with regular marijuana use.

A key assumption in the cumulative odds model is that explanatory effects will appear consistent across each level of a response variable (Long 1997, 140).[2] The assumption is often referred to as "parallel lines" or "parallel slopes." In the preceding case, odds ratios of 2.1779, 2.112, and 1.8184 did not differ significantly from one another, thus justifying the use of ordinal regression analysis. Had significant differences emerged across levels of the response measure, then a multinomial regression analysis might have been a more appropriate technique. Multinomial analyses do not offer the parsimony that ordinal regression models do (see Campbell and Donner 1989), but they do inform instances of departure from statistical assumptions.

Simple Ordinal Logistic Regression Analysis

Table 8.3 contains a simple ordinal regression model based on the data included in Tables 8.1 and 8.2. Threshold values are equivalent to intercepts (see Norusis 2005, 73), and Location values offer information about explanatory variables. As with binary and multinomial logistic regression models, the results of an ordinal regression analysis include parameter estimates and accompanying standard errors, Wald values and accompanying degrees of freedom, as well as significance tests. In Table 8.3, the parameter estimate for males is – .669. In binary and multinomial logistic regression analyses, one would exponentiate – .669 to arrive at an odds ratio of .512. But the ordinal regression model fit through the SPSS PLUM procedure estimates the cumulative logit by *subtracting* parameter estimates from appropriate thresholds (O'Connell 2006, 26).[3] Thus, the negative of – .669 is exponentiated, equaling 1.952. Consistent with the calculations from Table 8.2, one would conclude that, on average, males were approximately two times as likely as females to estimate a lower level of risk associated with regular marijuana use. This difference was significant at $p < .001$.

As with logistic regression models discussed in Chapters 6 and 7, ordinal regression procedures use –2 log likelihood values to indicate whether a model

Table 8.3 Ordinal logistic regression model testing sex as a predictor of marijuana risk perceptions

		B	SE	Wald	df	Sig.
Threshold	No Risk	−2.219	.078	804.443	1	.000
	Slight Risk	−1.050	.063	275.282	1	.000
	Moderate Risk	−.005	.059	.007	1	.931
Location	Males	−.669	.079	71.006	1	.000

containing explanatory measures offers a better fit than a model containing the intercept only. In the present analysis, a model containing the intercept only showed a –2 log likelihood value of 116.477, and the final model showed a value of 44.577. The chi-square difference, 71.900 with 1 degree of freedom, showed significance at $p < .001$, indicating a better fit for the model containing the sex determinant. In addition to log-likelihood values, the ordinal regression procedure offers a goodness-of-fit statistic, with a non-significant chi-square value indicating an acceptable fit to the data. In the case of sex and marijuana risk perceptions, the test showed a chi-square value of 3.540 with 2 degrees of freedom and a p-value of .170, indicating an acceptable fit. For the data in Table 8.3, three pseudo R-square measures ranged from .013 to .035, and the test of parallel lines (i.e., proportional odds) showed a non-significant p-value, thus satisfying the assumption.

Multiple Ordinal Logistic Regression Analysis

Table 8.4 expands the information from Table 8.3 to include three measures – sex, age, and the extent to which teachers had discussed the dangers of illicit substances – as determinants of marijuana risk perceptions. In this ordinal regression model, sex remained a significant predictor of perceived risk, but a dichotomous age indicator did not show significance. The ordinal variable measuring efforts of teachers to communicate drug risks began with Not Vigorous and concluded with Very Vigorous. Treated as a categorical factor, the measure showed significance at each level. The odds of an individual who indicated Not Vigorous perceiving a lower level of marijuana risk were 1.98 times the same odds of an individual who indicated Very Vigorous. The odds of a respondent who indicated Slightly Vigorous estimating a lower level of marijuana risk were 1.91 times the same odds for an individual indicating Very Vigorous communication. Thus, it appears that teacher efforts to communicate the hazards of drug use related significantly to adolescent estimates of the risks of regular marijuana use.

Research on Vague Quantifiers and Ordinal Measurement

"Ordinal data are the most common form of data acquired in the social sciences," Johnson and Albert (1999, v) explained, "but the analysis of such data is generally performed without regard to their ordinal nature." That observation certainly holds true in communication research, where ordinal measures are plentiful. As an example, studying the influence of television news on voter attitudes during election campaigns, survey researchers frequently ask respondents to estimate attention to network news. In the American National Election Studies, response options typically include No Attention, Very Little Attention, Some Attention, Quite a Bit of Attention, or A Great Deal of Attention. Inquiring about political leanings, pollsters may ask voters to indicate Very Liberal, Somewhat Liberal, Moderate, Somewhat Conservative, or Very Conservative. In a study addressing the health risks of alcohol use, research participants may estimate Great Risk, Moderate Risk, Slight Risk, or No Risk. In each of these situations, a series of "vague quantifiers" serve as ordered response options.

Vague quantifiers allow individuals to estimate attitudes and behaviors in a conversational manner while also allowing researchers to quantify responses systematically. But how should these response options be measured? Can equal intervals between observation points be assumed, or is ordinal measurement the only acceptable approach? Since the point at which C. I. Mosier (1941) observed contextual effects on the meaning that individuals ascribed to attitudinal terms, scholars have discussed and debated measurement strategies for vague quantifiers and ordinal variables (see Borgers, Hox, and Sikkel 2003, Bradburn and Miles 1979, Goocher 1965, Griffin 2013, Hakel 1969, Pepper and Prytulak 1974, Pracejus, Olsen, and Brown 2003, Schaeffer 1991, Wanke 2002, Wright, Gaskell, and O'Muircheartaigh 1994).

Generally, when a dependent variable contains at least four categories, with a normal distribution of error terms and equal variances, a "quasi-interval" approach is acceptable (DeMaris 1992, 77). Such an approach facilitates the use of ordinary least squares regression and the analysis of variance. But vague quantifiers can present methodological challenges, due primarily to inconsistencies with interpretation. Kennamer (1992) observed differences in the meaning individuals assigned to terms such as Often, Sometimes, Rarely, and Never, while Schaeffer (1991) found that meaning ascribed to vague terms varied by race, education level, and age. Schwarz, Grayson, and Knauper (1998) found that respondents drew on the formal features of a questionnaire in assigning meaning to vague terms, while Wanke (2002) found that respondents constructed meaning based on who they perceived as a target audience for a survey. Wright, Gaskell, and O'Muircheartaigh (1994) noted that meaning ascribed to vague quantifiers depended on the experiences of those making the assignments as well as the close peer groups of those individuals.

Relatedly, observing research that had identified associations between media use and media credibility, Rimmer and Weaver (1987) found that such associations depended on the operationalization of media use. Some surveys included affective attitudinal questions, the authors noted, while others asked respondents to quantify specific behaviors. The authors found that attitudinal items correlated more consistently with media credibility, suggesting potential differences in numeric and semantic response options (see also Baghal 2011, Eveland, Hutchens, and Shen 2009). As Wanke (2002, 301) thus summarized,

"(P)eople with objectively equal behaviour frequencies may indicate different scale values and people giving equal responses may actually differ in their objective behaviour frequencies."

A key point in this discussion is that when researchers use techniques such as ordinary least squares regression in analyses containing vague quantifiers, they make the implicit assumption that respondents assign the same meaning to ordered terms (Daykin and Moffatt 2002, 159). As the research cited above indicates, that assumption can be risky. One strength of the ordinal logistic regression model is that it allows for variation in meaning and interpretation of survey items (Anderson 1984). The model is thus uniquely suited for analyses involving vague quantifiers.

When working with vague quantifiers and discrete ordinal units, researchers might consider the following points in selecting a regression technique:

- Examine descriptive statistics and the distribution of scores. If a response measure is skewed, ordinal regression or a nonparametric ranks technique may perform better than ordinary least squares. Nussbaum (2015, 129–163) offers helpful strategies for selecting nonparametric tests, in particular.
- Note the number of categories in the response measure. If it contains fewer than four categories, an ordinal or multinomial regression procedure should be used ahead of ordinary least squares.
- Consider the nature of the research question. Will ordinal regression answer the question as well ordinary least squares?
- Compare the output of ordinary least squares and ordinal logistic regression. If marked differences occur, such that substantive conclusions stand to be affected, which output appears more reasonable given the data and the assumptions of the statistical procedures?

In the ordinal regression analysis displayed in Table 8.4, a model containing the intercept only showed a –2 log likelihood value of 421.010, and the final model showed a value of 312.241. The chi-square difference, 108.770 with 6 degrees of freedom, showed significance at $p < .001$, indicating a better fit for the model containing the three determinants. The goodness-of-fit test showed a chi-square value of 53.902 with 51 degrees of freedom and a p-value of .364, indicating an acceptable fit. Three pseudo R-square measures ranged from .023 to .063, but the multiple regression analysis did not meet the assumption of proportional odds. The null hypothesis – that slope coefficients would be the same across each level of the response measure – showed a –2 log likelihood value of 312.241. The "General" model showed a value of 290.887, revealing a chi-square difference of 21.354 with 12 degrees of freedom and a p-value of .045.

Given the lack of parallel slopes, a multinomial logistic regression model tested the explanatory effects of sex, age, and teacher communication on perceived risk of regular marijuana use. Table 8.5 contains the results of the multinomial analysis, and examining the exponentiated parameter estimates for

Table 8.4 Multiple ordinal logistic regression model testing sex, age, and teacher communication about drugs as determinants of marijuana risk perceptions

		B	*SE*	*Wald*	*df*	*Sig.*
Threshold	No Risk	–2.704	.124	474.455	1	.000
	Slight Risk	–1.443	.110	171.550	1	.000
	Moderate Risk	–.352	.105	11.245	1	.001
Location	Males	–.776	.088	77.523	1	.000
	Less than 18	–.072	.088	.678	1	.410
	Not Vigorous	–.683	.142	23.027	1	.000
	Slightly Vigorous	–.646	.137	22.238	1	.000
	Somewhat Vigorous	–.293	.128	5.196	1	.023
	Fairly Vigorous	–.243	.123	3.909	1	.048

Table 8.5 Multiple multinomial logistic regression model testing sex, age, and teacher communication about drugs as determinants of marijuana risk perceptions

Response Categories	*Explanatory Categories*	*B*	*SE*	*Wald*	*df*	*Sig.*	*Exp(B)*
No Risk	Intercept	–2.257	.200	127.758	1	.000	
	Males	1.209	.162	55.904	1	.000	3.351
	Less than 18	.148	.156	.903	1	.342	1.159
	Not Vigorous	1.038	.232	20.035	1	.000	2.823
	Slightly Vigorous	.886	.237	13.980	1	.000	2.426
	Somewhat Vigorous	.301	.235	1.642	1	.200	1.351
	Fairly Vigorous	–.029	.239	.014	1	.904	.972
Slight Risk	Intercept	–1.544	.160	92.924	1	.000	
	Males	.848	.132	41.521	1	.000	2.335
	Less than 18	.057	.131	.190	1	.663	1.059
	Not Vigorous	.533	.218	5.980	1	.014	1.704
	Slightly Vigorous	.764	.208	13.543	1	.000	2.146
	Somewhat Vigorous	.376	.195	3.706	1	.054	1.456
	Fairly Vigorous	.526	.181	8.446	1	.004	1.691
Moderate Risk	Intercept	–.941	.138	46.363	1	.000	
	Males	.356	.121	8.665	1	.003	1.427
	Less than 18	.141	.121	1.368	1	.242	1.151
	Not Vigorous	.129	.209	.384	1	.536	1.138
	Slightly Vigorous	.509	.192	7.023	1	.008	1.664
	Somewhat Vigorous	.362	.172	4.465	1	.035	1.437
	Fairly Vigorous	.281	.166	2.891	1	.089	1.325

the sex measure, it appears odds ratios for males and females might have differed across *no risk, slight risk,* and *moderate risk.* In the simple regression model, which contained the sex determinant only, differences across slopes did not emerge, but with two other explanatory variables present, relationships appeared to change. Given the reference category Great Risk, the odds of a male indicating no risk were 3.351 times the odds for females. Odds dropped to 2.335 at the level of Slight Risk and 1.427 in the category of Moderate Risk. Differences thus appeared most pronounced at No Risk.

Consistent with the ordinal regression analysis, age did not show significance as an explanatory measure, and it appears the strongest odds ratios for teacher communication occurred at the No Risk level of the dependent variable. With Great Risk as the reference category, the odds of an individual who responded Not Vigorous estimating No Risk were nearly three times the same odds of a respondent indicting Very Vigorous. At 2.426, the odds ratio remained strong at the Slightly Vigorous level of the explanatory variable, before losing significance at Somewhat Vigorous and Fairly Vigorous. At the Slight Risk level of the response measure, odds ratios did not appear as notable, but all four levels of the explanatory variable showed significance. At the third level of the dependent measure, Moderate Risk, relationships appeared comparatively weak.

So what should one make of the differences across results shown in the ordinal and multinomial logistic regression models? In this case, differences were not major, but the absence of parallel slopes did reveal potentially important information at the No Risk level of the dependent variable. Perhaps research in risk perceptions would reveal disparities among males and females at this level. Additionally, individuals who reported little communication on the part of teachers appeared especially likely to estimate No Risk associated with regular marijuana use. But beyond those patterns, a researcher would want to be careful about identifying differences across the two models. As O'Connell (2006, 29) explained, the test used to measure parallel slopes is not an especially powerful one and in the presence of multiple explanatory measures and/or large datasets, it might show statistically significant differences that have few practical implications. In general, a researcher would want to report statistical results in a conservative manner, identifying consistencies and inconsistencies in the models and with the theoretical framework guiding the analysis.

Interactions

As indicated in previous chapters, an interaction occurs when the effects of an explanatory measure on a response variable are influenced by a third measure, termed a *moderator* (Jaccard 2001). Interactions were tested in the logistic regression models discussed in the previous paragraphs, but significant effects did not emerge. The important point, for purposes of this chapter, is that researchers can analyze interactions in equations with ordinal response measures. In some cases, depending on conceptual development, hypotheses containing interactions can be tested.

Analyzing Source Attribution with Ordinal Measures

Sources define the news for mass audiences, Sigal (1973) suggested, and in the United States, public officials tend to dominate the conversation. As indicated in Chapter 6, journalists at prominent news agencies occassionally grant their sources anonymity, encouraging the sources to share information they otherwise might not. While anonymous attribution can protect news sources from retaliation and is generally considered a necessary practice in journalism, excessive use of it can lead to exaggerations, inaccuracies, and personal attacks on political foes. It therefore is important to monitor veiled sourcing in the news.

Denham (1997) conducted a study of anonymous sourcing in the Associated Press, the *Los Angeles Times* and the *Washington Post*, focusing on coverage of the military conflicts in Bosnia and Somalia from 1992 to 1994. The coding procedure in that study facilitated the use of ordinal logistic regression, and quantitative researchers may find the procedure useful for sourcing analyses in general. For each news report ($N = 472$), the author created a fraction in which the number of paragraphs citing a veiled source served as the numerator and the total number of paragraphs served as the denominator. The resulting decimal value was then rounded to lesser of two whole numbers. As an example, if a 27-paragraph article contained 6 paragraphs citing an anonymous source, the fraction $6/27 = .22 = 2$. If a 30-paragraph article contained 10 paragraphs citing an anonymous source, $10/30 = .33 = 3$. Every article in the study thus received a discrete ordered value reflecting its use of anonymous attribution.

The study found that, in coverage of Bosnia and Somalia, anonymous attribution varied across news organizations, suggesting editorial discretion, as opposed to the dynamics of individual news events, as a conceptual determinant of veiled sourcing. The Associated Press appeared the most conservative in citing anonymous sources while the *Washington Post* used veiled sources more frequently. In terms of analytic assumptions, the coding procedure resolved issues with observation independence, as it used a whole number derived from paragraphs in an article to represent the presence of anonymous attribution. Paragraphs themselves could not be considered independent of one another, and therefore using the paragraph as a unit of measurement would not have been appropriate. Yet, coding news reports for the mere presence of a veiled source may have glossed over their use. For instance, while 75% of articles on a given topic may cite an anonymous source, 85% of *all* sources may be identified by name. Technically, Denham (1997) used news articles as the unit of measurement, but paragraphs within the articles indicated the extent to which veiled sources appeared in the news. The current chapter includes an exercise on this sourcing measure, enhancing the approach taken in Chapter 6.

Ordinal Logistic Regression in SPSS

This section of the chapter offers instruction on how to conduct an ordinal logistic regression analysis in SPSS. In doing so, the chapter draws on three categorical explanatory variables and one ordinal response measure from the 2008 American National Election Study (The American National Election Studies 2008). Serving as a response measure was a discrete interval variable indicating the number of days (0 to 7) in a typical week respondents had discussed politics with family members or friends. In this analysis, because the category indicating six days per week contained very few observations, it was combined with five days per week. Explanatory factors consisted of a sex measure, a three-category variable indicating party identification (Democrat, Republican, Independent), and a five-category variable measuring attention to politics on television news. ANES researchers introduced the attention variable, which included response options *a great deal, a lot, a moderate amount, a little,* and *none,* as a new measure in 2008, and only a portion of study participants were asked to respond to it. Because very few respondents indicated paying *no* attention, it was combined with *a little.* Analyses in this section are based on responses from 888 individuals.

Figure 8.1 contains screenshots for (a) beginning an ordinal regression analysis in SPSS and (b) selecting output options in the SPSS PLUM procedure. To begin an ordinal regression, a researcher should first select Analyze, followed by Regression, followed by Ordinal. At that point, a window will open allowing the researcher to enter explanatory and response measures. Categorical predictors should be entered as Factors and continuous determinants as Covariates. This is an important step in ordinal regression analysis, as continuous measures can increase the number of cells rapidly and potentially cause zero-count cells to emerge. While the ordinal regression procedure handles continuous measures effectively, the procedure does require a sufficiently large sample in the presence of metric covariates.

The same concern arises with nominal and ordinal predictors as well as ordinal response measures. If a given category contains few observations, empty cells will emerge quickly in a multivariate analysis, compromising otherwise stable parameter estimates. As with the multinomial model, a researcher can specify a value for Delta in the Options window of the SPSS ordinal regression procedure. Delta should be used only when empty cells appear, its value set between 0 and 1.[4] Linking theory with method, a researcher should also consider *why* a category contains few observations. Does the category need to be present? What might be gained and what might be lost in collapsing categories?

The second window in Figure 8.1 contains output options for the ordinal regression procedure. As shown in the figure, multiple options exist, and the checked boxes indicate those selected for the current example. These options include goodness-of-fit statistics, summary statistics, parameter estimates, cell

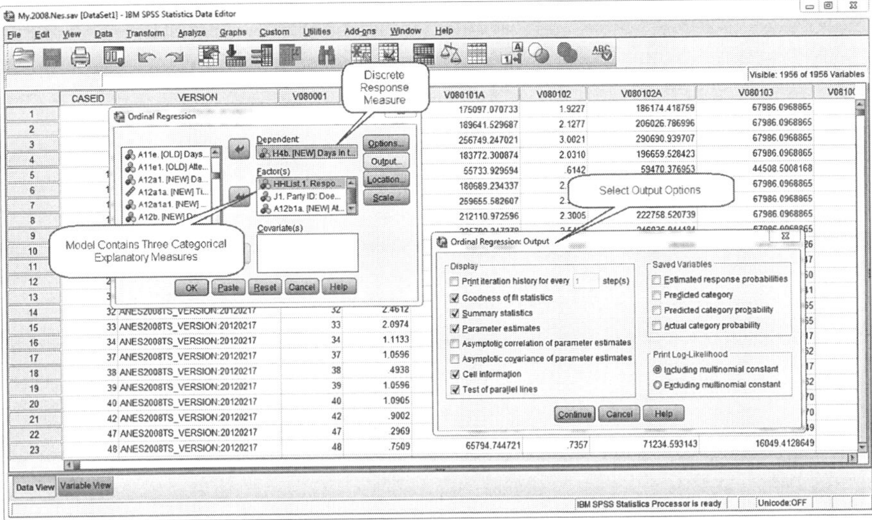

Figure 8.1 SPSS screenshots for output options in ordinal logistic regression analysis (PLUM). Source: SPSS® Reprints Courtesy of International Business Machines Corporation, © 2014 International Business Machines Corporation

information, and a test of parallel lines. Output will be reviewed after additional SPSS windows are discussed.

Figure 8.2 contains Location options in the ordinal regression procedure. The current example includes three categorical predictors and no interaction terms, thus analyzing main effects only. Prior to the main-effects model, analyses included interactions, and as shown in the Location window, SPSS allows the researcher to build terms into the model in a specific order. One can test a single two-way interaction in addition to main effects, or one can perform a hierarchical analysis including all interactions, provided the researcher has a theoretical reason for testing such a model. Norusis (2005) offers an excellent discussion of the SPSS ordinal logistic regression procedure, while Stokes, Davis, and Koch (2012) provide detailed information for analyses in SAS.

Table 8.6 contains statistical output corresponding to the SPSS displays in Figures 8.1 and 8.2. The first set of statistics contains –2 log likelihood values for the intercept-only and final models. As shown in the table, the chi-square difference between the two, 101.203 with 6 degrees of freedom, proved significant, allowing one to conclude an improved fit for the final model. Next, goodness-of-fit statistics did not show significance, indicating a satisfactory fit, and Pseudo R-square measures ranged from .031 to .111. Examining parameter estimates, neither sex nor political party predicted the number of days in a typical week that respondents discussed politics. However, the parameter estimate (1.692) indicated

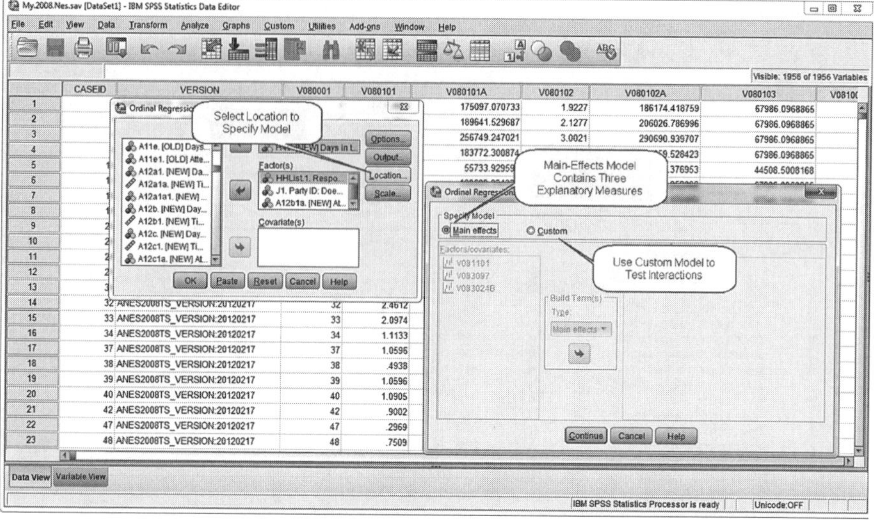

Figure 8.2 SPSS screenshots for location options in ordinal logistic regression analysis (PLUM). Source: SPSS® Reprints Courtesy of International Business Machines Corporation, © 2014 International Business Machines Corporation

that the odds of those who paid a great deal of attention to TV news discussing politics comparably few days per week were just .184 times the odds of those who paid little or no attention to TV news. Expontiating the negative of 1.326, one would conclude that odds of those who paid a lot of attention to TV news indicating fewer days discussing politics were .266 times the odds of those who paid little or no attention. Thus, one observes a pattern in which those who paid more attention to TV news discussed politics more frequently. Continuing with Table 8.6, the test of parallel lines did not show significance at $p < .05$, and therefore one may conclude that an ordinal regression model was the appropriate choice.[5] Had the results shown significance, one could not have concluded proportionate odds, and a multinomial model might have helped to clarify relationships.

Lastly, Tables 8.7a and 8.7b contain observed and expected cell frequencies as well as Pearson residual values reflecting their differences. As indicated earlier in the table, the model fit the data in a statistically acceptable manner, and to some extent, the residuals, which Norusis (2005, 77) explained reflect standardized differences between observed and expected values, illustrate that fit.

Chapter Summary

This chapter has focused on ordinal logistic regression analysis, a technique used when a response measure contains a series of ordered categories. As with other logistic regression models, ordinal regression produces parameter estimates in

Table 8.6 SPSS output for ordinal logistic regression model containing categorical predictors

	Model Fitting Information			
	Model Fitting Criteria	Likelihood Ratio Tests		
Model	*−2 Log Likelihood*	*Chi-Square*	*df*	*Sig.*
Intercept Only	630.247			
Final	529.045	101.203	6	.000

	Goodness-of-Fit		
	Chi-Square	*df*	*Sig.*
Pearson	130.696	132	.516
Deviance	136.058	132	.387

Pseudo R-Square

Cox and Snell	.108
Nagelkerke	.111
McFadden	.031

Parameter Estimates

	B	SE	Wald	df	Sig.	95% Confidence Interval Lower	Upper
Threshold							
Discuss Politics 0 Days / Week	−.558	.155	12.946	1	.000	−.861	−.354
Discuss Politics 1 Day / Week	.643	.155	17.197	1	.000	.339	.947
Discuss Politics 2 Days / Week	1.405	.161	76.555	1	.000	1.090	1.719
Discuss Politics 3 Days / Week	2.192	.169	167.713	1	.000	1.860	2.523
Discuss Politics 4 Days / Week	2.511	.174	208.296	1	.000	2.170	2.851
Discuss Politics 5/6 Days / Week	3.129	.186	281.501	1	.000	2.763	3.494

(Continued)

Table 8.6   (Continued)

		Parameter Estimates					
							95% Confidence Interval
	B	*SE*	*Wald*	*df*	*Sig.*	*Lower*	*Upper*
Location							
Male	.019	.121	.025	1	.874	−.218	.257
Democrat	.086	.140	.375	1	.540	−.189	.361
Republican	.121	.167	.528	1	.467	−.206	.449
A Great Deal of Attention	1.692	.194	76.136	1	.000	1.312	2.073
A Lot of Attention	1.326	.180	54.530	1	.000	.974	1.678
A Moderate Amount of Attention	.694	.158	19.255	1	.000	.384	1.004

		Test of Parallel Lines [a]		
Model	−2 Log Likelihood	Chi-Square	df	Sig.
Null Hypothesis	529.045			
General	488.283	40.761	30	.091

The null hypothesis states that the location parameters (slope coefficients) are the same across response categories.

[a] Link function: Logit.

the form of log odds, and their exponentiation results in odds ratios. Ordinal regression assumes proportional odds across the categories of a response measure, and lacking proportionality, or parallel slopes, a multinomial model should be considered. The chapter also included information concerning the measurement of vague quantifiers.

Chapter Exercises

1. Chapter 6 contained an exercise on anonymous attribution in the news, as examined in Denham (2012). This question pertains to the same subject, except the two-level nominal measure of attribution used in Chapter 6 has been changed to an ordinal measure, as discussed in the sidebar "Analyzing Source Attribution with Ordinal Measures." Changes are reflected in the first column of each data grouping below. Use these data and an ordinal logistic regression analysis to examine the effects of time period, front-page placement, and dateline on the use of anonymous attribution in the news. In your analysis, report −2 log likelihood values for the Intercept Only and Final models, as well as the Pearson goodness-of-fit test and three R-squared

Table 8.7a   SPSS frequencies for ordinal logistic regression model containing categorical predictors

Sex	Party ID	Attention to TV News		Days in Typical Week Discuss Politics						
				0	1	2	3	4	5/6	7
Male	Democrat	A Great Deal	Observed	10	3	8	6	2	3	11
			Expected	3.726	6.578	7.025	8.354	3.172	5.160	8.985
			Residual	3.401	−1.516	.402	−.907	−.684	−1.014	.756
		A Lot	Observed	6	10	9	8	1	3	6
			Expected	5.177	8.263	7.776	8.089	2.790	4.251	6.655
			Residual	.386	.672	.485	−.035	−1.108	−.639	−.276
		A Moderate Amount	Observed	11	8	9	11	2	3	7
			Expected	10.441	13.067	9.481	7.860	2.347	3.280	4.524
			Residual	.194	−1.625	−.173	1.218	−.232	−.160	1.219
		A Little	Observed	10	7	5	1	0	3	1
			Expected	9.183	7.861	4.171	2.804	.746	.983	1.252
			Residual	.332	−.365	.442	−1.138	−.876	2.073	−.231
	Republican	A Great Deal	Observed	3	2	2	5	1	1	5
			Expected	1.594	2.838	3.064	3.690	1.414	2.317	4.082
			Residual	1.164	−.539	−.664	.760	−.362	−.923	.512
		A Lot	Observed	1	5	4	2	1	5	1
			Expected	2.217	3.578	3.411	3.595	1.251	1.918	3.029
			Residual	−.870	.835	.352	−.934	−.232	2.348	−1.267
		A Moderate Amount	Observed	5	10	6	2	2	2	5
			Expected	6.369	8.101	5.970	5.009	1.506	2.113	2.932
			Residual	−.606	.772	.014	−1.464	.412	−.081	1.267
		A Little	Observed	3	6	4	3	0	0	0
			Expected	5.315	4.653	2.508	1.701	.455	.601	.767
			Residual	−1.229	.742	1.026	1.053	−.684	−.790	−.898

(Continued)

Table 8.7a   (Continued)

			Days in Typical Week Discuss Politics						
Sex	Party ID	Attention to TV News	0	1	2	3	4	5/6	7
	Independent	A Great Deal	Observed						
			2	4	3	1	1	5	3
			Expected						
			1.780	3.077	3.196	3.685	1.367	2.188	3.707
			Residual						
			.173	.575	−.120	−1.558	−.326	2.021	−.409
		A Lot	Observed						
			1	3	6	5	1	3	3
			Expected						
			2.856	4.433	4.038	4.070	1.374	2.065	3.165
			Residual						
			−1.177	−.762	1.080	.511	−.329	.684	−.100
		A Moderate Amount	Observed						
			11	18	16	6	0	5	4
			Expected						
			13.144	15.797	11.032	8.885	2.608	3.612	4.921
			Residual						
			−.669	.646	1.656	−1.049	−1.651	.753	−.433
		A Little	Observed						
			15	8	5	3	3	1	2
			Expected						
			13.307	10,781	5.506	3.622	.954	1.249	1.580
			Residual						
			.580	−1.006	−.234	−.344	2.123	−.227	.341

Table 8.7b SPSS frequencies for ordinal logistic regression model containing categorical predictors

Sex	Party ID	Attention to TV News		Days in Typical Week Discuss Politics						
				Zero	One	Two	Three	Four	Five/Six	Seven
Female	Democrat	A Great Deal	Observed	4	5	5	6	3	6	12
			Expected	3.616	6.354	6.744	7.965	3.008	4.876	8.437
			Residual	.212	−.584	−.735	−.776	−.005	.542	1.376
		A Lot	Observed	7	13	8	10	6	7	7
			Expected	7.102	11.267	10.527	10.873	3.732	5.668	8.831
			Residual	−.041	.575	−.861	−.294	1.214	.589	−.669
		A Moderate Amount	Observed	13	27	12	16	4	2	9
			Expected	17.255	21.402	15.397	12.680	3.771	5.260	7.235
			Residual	−1.151	1.404	−.959	1.013	.121	−1.469	.687
		A Little	Observed	26	17	11	12	0	0	2
			Expected	23.423	19.808	10.419	6.970	1.851	2.434	3.096
			Residual	.658	−.750	.196	2.011	−1.379	−1.589	−.637
	Republican	A Great Deal	Observed	1	6	5	2	2	2	3
			Expected	1.793	3.178	3.411	4.080	1.556	2.539	4.444
			Residual	−.619	1.719	.940	−1.147	.370	−.361	−.772
		A Lot	Observed	3	5	5	5	3	3	4
			Expected	3.324	5.331	5.047	5.282	1.829	2.794	4.393
			Residual	−.189	−.159	−.023	−.136	.895	.130	−.204
		A Moderate Amount	Observed	8	15	4	5	2	2	5
			Expected	8.287	10.448	7.635	6.364	1.906	2.669	3.691
			Residual	−.112	1.631	−1.458	−.588	.070	−.423	.714
		A Little	Observed	6	5	7	2	1	1	0
			Expected	7.403	6.402	3.421	2.310	.616	.812	1.036
			Residual	−.633	−.658	2.106	−.215	.496	.212	−1.043

(*Continued*)

Table 8.7b   (Continued)

Sex	Party ID	Attention to TV News		Days in Typical Week Discuss Politics						
				Zero	One	Two	Three	Four	Five/Six	Seven
	Independent	A Great Deal	Observed	0	1	1	5	3	1	2
			Expected	1.240	2.132	2.200	2.519	.930	1.483	2.497
			Residual	−1.171	−.848	−.888	1.741	2.228	−.421	−.350
		A Lot	Observed	4	4	2	5	2	4	3
			Expected	3.168	4.886	4.418	4.421	1.485	2.226	3.396
			Residual	.502	−.449	−1.274	.305	.436	1.249	−.232
		A Moderate Amount	Observed	12	12	8	5	1	1	3
			Expected	9.340	11.121	7.700	6.162	1.802	2.491	3.384
			Residual	.987	.307	.120	−.507	−.611	−.974	−.218
		A Little	Observed	21	17	9	9	2	1	0
			Expected	21.482	17.187	8.703	5.699	1.497	1.958	2.474
			Residual	−.130	−.054	.109	1.455	.417	−.696	−1.607

estimates. Also report parameter estimates and their standard errors, in addition to Wald values, significance tests, exponentiated parameter estimates, and 95% confidence intervals. Lastly, test for parallel lines in the model. Are the results of the ordinal analysis consistent with those reported in Chapter 6?

ID	A P F D O	ID	A P F D O	ID	A P F D O	ID	A P F D O	ID	A P F D O
001	0 1 2 1 0	034	0 1 2 1 0	067	2 2 2 2 0	100	1 2 2 2 0	133	0 1 2 2 0
002	1 1 1 1 0	035	1 1 2 1 1	068	2 2 2 1 0	101	1 2 2 1 0	134	0 1 2 2 1
003	0 1 2 1 0	036	0 1 2 1 1	069	1 2 2 2 0	102	1 2 2 1 0	135	3 1 1 1 0
004	1 1 1 1 0	037	0 1 2 2 0	070	1 2 2 1 0	103	1 2 2 1 0		
005	0 1 2 1 1	038	1 1 2 1 0	071	2 2 2 1 0	104	1 2 2 1 0		
006	1 1 1 1 1	039	0 1 2 1 0	072	1 2 1 1 0	105	1 1 2 1 0		
007	0 1 2 1 0	040	2 1 2 1 0	073	0 2 2 1 0	106	0 1 2 1 0		
008	1 1 2 2 0	041	0 1 2 1 0	074	1 2 2 2 0	107	0 1 2 1 0		
009	1 1 2 1 0	042	1 1 2 1 0	075	0 2 2 2 0	108	0 1 2 2 0		
010	0 1 2 1 0	043	1 1 2 1 1	076	0 2 2 1 0	109	0 1 2 1 0		
011	3 1 2 2 0	044	2 1 2 2 1	077	2 2 2 1 0	110	0 1 2 1 0		
012	0 1 2 1 0	045	1 1 2 2 1	078	0 2 2 1 0	111	0 1 2 1 0		
013	1 1 2 1 0	046	0 1 1 1 0	079	1 2 2 2 0	112	1 1 2 1 1		
014	3 1 2 2 0	047	1 1 1 1 1	080	1 2 1 1 0	113	1 1 2 2 0		
015	1 1 1 1 1	048	0 1 2 1 1	081	1 2 2 2 0	114	0 1 2 2 1		
016	3 1 1 2 0	049	0 1 2 1 1	082	1 2 2 2 0	115	1 1 1 2 0		
017	1 1 2 1 1	050	1 1 1 1 0	083	2 2 2 1 0	116	1 1 1 1 0		
018	0 1 2 1 0	051	1 1 1 1 1	084	3 2 2 1 0	117	1 1 2 1 0		
019	1 1 1 2 0	052	1 1 2 2 0	085	3 2 1 1 1	118	0 1 2 1 0		
020	3 1 1 2 1	053	0 1 2 1 0	086	3 2 2 1 0	119	3 1 2 1 0		
021	1 1 2 2 0	054	3 2 2 2 0	087	2 2 2 1 0	120	0 1 2 1 1		
022	3 1 1 2 1	055	0 2 2 1 0	088	0 2 2 1 0	121	0 1 2 1 1		
023	1 1 1 2 0	056	0 2 2 1 0	089	1 2 1 1 0	122	0 1 2 1 0		
024	1 1 1 1 1	057	1 2 1 1 0	090	1 2 2 2 0	123	1 1 2 2 0		
025	0 1 2 1 0	058	3 2 1 2 0	091	1 2 2 2 0	124	1 1 2 2 1		
026	0 1 2 1 0	059	1 2 2 2 0	092	1 2 2 2 0	125	3 1 2 2 0		
027	2 2 2 1 0	060	0 2 1 1 0	093	1 2 2 2 1	126	0 1 2 1 1		
028	0 1 2 2 1	061	0 2 2 1 0	094	1 2 2 1 1	127	1 1 2 2 1		
029	1 1 1 2 0	062	0 2 1 2 0	095	0 2 2 2 0	128	0 1 2 1 0		
030	0 1 2 1 0	063	2 2 2 2 0	096	0 2 2 2 0	129	1 1 2 1 0		
031	0 1 2 1 0	064	0 2 2 1 1	097	1 2 1 1 0	130	2 1 2 2 0		
032	0 1 2 2 0	065	1 2 2 2 0	098	0 2 2 1 0	131	0 1 2 1 1		
033	1 1 2 2 0	066	0 2 2 1 0	099	1 2 2 2 0	132	0 1 2 1 0		

Note: Excel file containing data available on companion website.
Category codes:

Anonymous attribution (A):	Ordered values, 0 to 3
Period (P):	1 = Before 9/11, 2 = After 9/11
Front page news report (F):	1 = Yes, 2 = No
Dateline (D):	1 = US, 2 = International
Office of National Drug Control Policy source present (O):	0 = No, 1 = Yes

2. Organizational practitioners frequently conduct market research, and this question contains information about interest in professional baseball among members of racial minorities, most of whom are African American. Interest in baseball is measured on three levels (Very Interested, Somewhat Interested, Not at all Interested), and additional measures include sex as well as level of education (High School Graduate or Less, Some College, College Graduate, Graduate School). Given these data, gathered in a CBS News poll (CBS News 2008), use an ordinal logistic regression analysis to examine the effects of sex and education level on interest in professional baseball. In your analysis, report −2 log likelihood values for the Intercept Only and Final models, as well as the Pearson goodness-of-fit test and three R-squared estimates. Also report parameter estimates and their standard errors, in addition to Wald values, significance tests, exponentiated parameter estimates, and 95% confidence intervals. Lastly, test for parallel lines in the model. What can one conclude about interest in baseball given the determinants of sex and level of education?

ID	B S E	ID	B S E	ID	B S E	ID	B S E	ID	B S E	ID	B S E	ID	B S E
001	1 1 1	026	2 1 1	051	2 2 1	076	2 2 4	101	3 1 4	126	3 2 1	151	3 2 2
002	1 1 1	027	2 1 1	052	2 2 1	077	2 2 4	102	3 1 4	127	3 2 1	152	3 2 2
003	1 1 1	028	2 1 1	053	2 2 1	078	3 1 1	103	3 1 4	128	3 2 1	153	3 2 2
004	1 1 1	029	2 1 1	054	2 2 1	079	3 1 1	104	3 1 4	129	3 2 1	154	3 2 2
005	1 1 2	030	2 1 1	055	2 2 1	080	3 1 1	105	3 1 4	130	3 2 1	155	3 2 2
006	1 1 2	031	2 1 1	056	2 2 1	081	3 1 1	106	3 1 4	131	3 2 2	156	3 2 2
007	1 1 3	032	2 1 1	057	2 2 1	082	3 1 1	107	3 1 4	132	3 2 2	157	3 2 2
008	1 1 3	033	2 1 1	058	2 2 1	083	3 1 1	108	3 1 4	133	3 2 2	158	3 2 2
009	1 1 3	034	2 1 2	059	2 2 1	084	3 1 1	109	3 1 4	134	3 2 2	159	3 2 3
010	1 1 4	035	2 1 2	060	2 2 1	085	3 1 1	110	3 1 4	135	3 2 2	160	3 2 3
011	1 1 4	036	2 1 2	061	2 2 2	086	3 1 1	111	3 2 1	136	3 2 2	161	3 2 3
012	1 2 1	037	2 1 2	062	2 2 2	087	3 1 1	112	3 2 1	137	3 2 2	162	3 2 3
013	1 2 1	038	2 1 2	063	2 2 2	088	3 1 2	113	3 2 1	138	3 2 2	163	3 2 3
014	1 2 1	039	2 1 2	064	2 2 2	089	3 1 2	114	3 2 1	139	3 2 2	164	3 2 3
015	1 2 1	040	2 1 2	065	2 2 2	090	3 1 2	115	3 2 1	140	3 2 2	165	3 2 4
016	1 2 1	041	2 1 2	066	2 2 2	091	3 1 2	116	3 2 1	141	3 2 2	166	3 2 4
017	1 2 2	042	2 1 3	067	2 2 3	092	3 1 2	117	3 2 1	142	3 2 2	167	3 2 4
018	1 2 2	043	2 1 3	068	2 2 3	093	3 1 2	118	3 2 1	143	3 2 2	168	3 2 4
019	1 2 3	044	2 1 3	069	2 2 3	094	3 1 2	119	3 2 1	144	3 2 2	169	3 2 4
020	1 2 3	045	2 1 4	070	2 2 3	095	3 1 2	120	3 2 1	145	3 2 2	170	3 2 4
021	1 2 3	046	2 1 4	071	2 2 3	096	3 1 2	121	3 2 1	146	3 2 2	171	3 2 4
022	1 2 3	047	2 1 4	072	2 2 3	097	3 1 3	122	3 2 1	147	3 2 2	172	3 2 4
023	1 2 3	048	2 1 4	073	2 2 3	098	3 1 3	123	3 2 1	148	3 2 2	173	3 2 4
024	1 2 4	049	2 1 4	074	2 2 3	099	3 1 3	124	3 2 1	149	3 2 2	174	3 2 4
025	2 1 1	050	2 1 4	075	2 2 4	100	3 1 3	125	3 2 1	150	3 2 2	175	3 2 4

Note: Excel file containing data available on companion website.

Category codes:

Interest in Baseball (B):   1 = Very Interested, 2 = Somewhat Interested, 3 = Not at all Interested
Sex (S):   1 = Males, 2 = Females
Level of Education (E):   1 = High School or Less, 2 = Some College, 3 = College Graduate,
4 = Graduate School

Notes

1   While the cumulative odds model indicates the odds of an observation appearing at or below a specific level of the response measure, the adjacent categories model compares each response category with the next largest. Continuation-ratio models compare a response category with all lower categories. For additional discussion, Hosmer and Lemeshow (2000, 288–308) discuss the adjacent categories and continuation-ratio models (see also Agresti 1984, 113–114, O'Connell 2006, 54–84, Powers and Xie 2000, 201–222).

2   Borooah (2002, 6) explained the parallel slopes assumption with a practical example: "(I)f there is a variable which affects the likelihood of a person being in the ordered categories (e.g., diet on health status), then it is assumed that the coefficients linking the variable value to the different outcomes will be the same across all the outcomes (a given diet will affect the likelihood of a person being in excellent health exactly the same as it will affect the likelihood of him or her being in poor health)."

3   Commenting on the subtraction process, Norusis (2005, 71) explained, "That is done so that larger coefficients indicate an association with larger scores. When you see a positive coefficient for a dichotomous factor, you know that higher scores are more likely for the first category. A negative score tells you that lower scores are more likely." The SPSS approach thus helps to reconcile output across the three types of logistic regression analyses.

4   A choice of link functions also appears in the Options window. In addition to the logit, which serves as an effective link in the majority of cases, additional link functions include probit, addressed in Chapter 9 of the current text, as well as complementary log-log, negative log-log, and Cauchit. Norusis (2005, 84) explained that complementary log-log is appropriate when the higher categories of a measure are more probable, while negative log-log may prove useful when lower categories are more probable. Cauchit, Norusis explained, is useful when an outcome contains many extreme values.

5   Norusis (2005, 74) explained the test for parallel slopes in more specific detail: "The row labeled *Null Hypothesis* contains –2 log-likelihood for the constrained model, the model that assumes the lines are parallel. The row labeled *General* is for the model with separate lines or planes. You want to know whether the general model results in a sizeable improvement in fit from the null hypothesis model."

References

Agresti, Alan. 1984. *Analysis of Ordinal Categorical Data*. New York: John Wiley & Sons.
Agresti, Alan. 1989. "Tutorial on Modeling Ordered Categorical Response Data." *Psychological Bulletin*, 105: 290–301. DOI:10.1037/0033–2909.105.2.290.

Ananth, Cande V., and David G. Kleinbaum. 1997. "Regression Models for Ordinal Responses: A Review of Methods and Applications." *International Journal of Epidemiology*, 26: 1323–1333. DOI:10.1093/ije/26.6.1323.

Anderson, J. A. 1984. "Regression and Ordered Categorical Variables." *Journal of the Royal Statistical Society*, 46: 1–30.

Baghal, Tarek. 2011. "The Measurement of Risk Perceptions: The Case of Smoking." *Journal of Risk Research*, 14: 351–364. DOI:10.1080/13669877.2010.541559.

Borgers, Natacha, Joop Hox, and Dirk Sikkel. 2003. "Response Quality in Survey Research with Children and Adolescents: The Effect of Labeled Response Options and Vague Quantifiers." *International Journal of Public Opinion Research*, 15: 83–94.

Borooah, Vani K. 2002. *Logit and Probit: Ordered and Multinomial Models*. Thousand Oaks, CA: Sage.

Bradburn, Norman M., and Carrie Miles. 1979. "Vague Quantifiers." *Public Opinion Quarterly*, 43: 92–101. DOI:10.1086/268494.

Campbell, M. Karen, and Allan Donner. 1989. "Classification Efficiency of Multinomial Logistic Regression Relative to Ordinal Logistic Regression." *Journal of the American Statistical Association*, 84: 587–591. DOI:10.1080/01621459.1989.10478807.

CBS News. 2008. CBS News Monthly Poll #1, March 2008. ICPSR26144-v1. Ann Arbor, MI: Inter-university Consortium for Political and Social Research [distributor], 2009-10-08. DOI:10.3886/ICPSR26144.v1.

Chyi, Hsiang Iris, and Mengchieh Jacie Yang. 2009. "Is Online News an Inferior Good? Examining the Economic Nature of Online News among Users." *Journalism & Mass Communication Quarterly*, 86: 594–612. DOI:10.1177/107769900908600309.

Chyi, Hsiang Iris, Mengchieh Jacie Yang, Seth C. Lewis, and Nan Zheng. 2010. "Use of and Satisfaction with Newspaper Sites in the Local Market: Exploring Differences between Hybrid and Online-only Users." *Journalism & Mass Communication Quarterly*, 87: 62–83. DOI:10.1177/107769901008700104.

Clogg, Clifford C., and Edward S. Shihadeh. 1994. *Statistical Models for Ordinal Variables*. Thousand Oaks, CA: Sage.

Daykin, Anne R., and Peter G. Moffatt. 2002. "Analyzing Ordered Responses: A Review of the Ordered Probit Model." *Understanding Statistics*, 1: 157–166. DOI:10.1207/S15328031US0103_02.

DeMaris, Alfred. 1992. *Logit Modeling: Practical Applications*. Newbury Park, CA: Sage.

DeMaris, Alfred. 2004. *Regression with Social Data: Modeling Continuous and Limited Response Variables*. Hoboken, NJ: Wiley Interscience.

Denham, Bryan E. 1997. "Anonymous Attribution during Two Periods of Military Conflict: Using Logistic Regression to Study Veiled Sources in American Newspapers." *Journalism & Mass Communication Quarterly*, 74: 565–578. DOI:10.1177/107769909707400310.

Denham, Bryan E. 2009. "Determinants of Anabolic-Androgenic Steroid Risk Perceptions in Youth Populations: A Multivariate Analysis." *Journal of Health & Social Behavior*, 50: 277–292. DOI:10.1177/002214650905000303.

Denham, Bryan E. 2012. "NY Times War on Drugs Sources Change after September 11." *Newspaper Research Journal*, 33(4): 34–47.

Dunwoody, Sharon, Dominique Brossard, and Anthony Dudo. 2009. "Socialization or Rewards? Predicting U.S. Science-Media Interactions." *Journalism & Mass Communication Quarterly*, 86: 299–314. DOI:10.1177/107769900908600203.

Evans, W. Douglas, Kevin C. Davis, Olivia Silber Ashley, and Munziba Khan. 2012. "Effects of Media Messages on Parent–Child Sexual Communication." *Journal of Health Communication*, 17: 498–514. DOI:10.1080/10810730.2011.635772.

Eveland, William P., Myiah J. Hutchens, and Fei Shen. 2009. "Exposure, Attention, or 'Use' of News? Assessing Aspects of the Reliability and Validity of a Central Concept in Political Communication Research." *Communication Methods and Measures*, 3: 223–244. DOI: 10.1080/19312450903378925.

Goocher, Buell. 1965. "Effects of Attitude and Experience on the Selection of Frequency Adverbs." *Journal of Verbal Learning and Verbal Behavior*, 4: 193–195. DOI:10.1016/S0022-5371(65)80020-2 DOI:10.1016/S0022-5371%2865%2980020-2#doilink.

Griffin, Jamie. 2013. "On the Use of Latent Variable Models to Detect Differences in the Interpretation of Vague Quantifiers." *Public Opinion Quarterly*, 77: 124–144. DOI:10.1093/poq/nfs072.

Hakel, Milton D. 1969. "How Often is Often?" *American Psychologist*, 23: 533–534. DOI:10.1037/h0037716.

Han, Paul K. J., Richard Moser, and William M. P. Klein. 2006. "Perceived Ambiguity about Cancer Prevention Recommendations: Relationship to Perceptions of Cancer Preventability, Risk, and Worry." *Journal of Health Communication*, 11: 51–69. DOI: 10.1080/10810730600637541.

Hay, Jennifer, Elliot Coups, and Jennifer Ford. 2006. "Predictors of Perceived Risk for Colon Cancer in a National Probability Sample in the United States." *Journal of Health Communication*, 11: 71–92. DOI: 10.1080/10810730600637376.

Hayes, Andrew F. 2005. *Statistical Methods for Communication Science*. Mahwah, NJ: Erlbaum.

Hewes, Dean. 1978. "'Levels of Measurement' Problem in Communication Research: A Review, Critique, and Partial Solution." *Communication Research*, 5: 87–127. DOI:10.1177/009365027800500105.

Holtbrugge, Werner, and Martin Schumacher. 1991. "A Comparison of Regression Models for the Analysis of Ordered Categorical Data." *Applied Statistics*, 40: 249–259.

Hosmer, David W., and Stanley Lemeshow. 2000. *Applied Logistic Regression*. 2nd ed. New York: John Wiley & Sons.

Jaccard, James. 2001. *Interaction Effects in Logistic Regression*. Thousand Oaks, CA: Sage.

Johnson, Valen E., and James H. Albert. 1999. *Ordinal Data Modeling: Statistics for Social Science and Public Policy*. New York: Springer.

Johnston, Lloyd D., Jerald G. Bachman, Patrick M. O'Malley, and John Schulenberg. 2012. Monitoring the Future: A Continuing Study of American Youth. Funded by National Institute on Drug Abuse. Institutional for Social Research, University of Michigan.

Kennamer, J. David. 1992. "Use of 'Vague' Quantifiers in Measuring Communication Behaviors." *Journalism Quarterly*, 69: 646–650. DOI:10.1177/107769909206900312.

Long, J. Scott. 1997. *Regression Models for Categorical and Limited Dependent Variables.* Thousand Oaks, CA: Sage.

McCullagh, Peter. 1980. "Regression Models for Ordinal Data." *Journal of the Royal Statistical Society,* 42: 109–142.

Mosier, C. I. 1941. "A Psychometric Study of Meaning." *Journal of Social Psychology,* 13: 123–140. DOI:10.1080/00224545.1941.9714065.

Norman, Geoff. 2010. "Likert Scales, Levels of Measurement, and the 'Laws' of Statistics." *Advances in Health Sciences Education,* 15: 625–632. DOI:10.1007/s10459-010-9222-y.

Norusis, Marija J. 2005. *SPSS 14.0 Advanced Statistical Procedures Companion.* Upper Saddle River, NJ: Prentice Hall.

Nussbaum, E. Michael. 2015. *Categorical and Nonparametric Data Analysis.* New York: Routledge.

O'Connell, Ann A. 2006. *Logistic Regression Models for Ordinal Response Variables.* Thousand Oaks, CA: Sage.

Pepper, Susan, and Lubomir Prytulak. 1974. "Sometimes Frequently Means Seldom: Context Effects in the Interpretation of Quantitative Expressions." *Journal of Research in Personality,* 8: 95–101. DOI:10.1016/0092-6566(74)90049-X.

Powers, Daniel A., & Yu Xie. 2000. *Statistical Methods for Categorical Data Analysis.* San Diego, CA: Academic Press.

Pracejus, John W., G. Douglas Olsen, and Norman R. Brown. 2003. "On the Prevalence and Impact of Vague Quantifiers in the Advertising of Cause-Related Marketing (CRM)." *Journal of Advertising,* 32: 19–28. DOI:10.1080/00913367.2003.10639146.

Prochaska, J. O., and C. C. DiClemente. 1983. "Stages and Processes of Self-Change of Smoking: Toward an Integrative Model." *Journal of Counseling and Clinical Psychology,* 51: 390–395.

Rimmer, Tony, and David Weaver. 1987. "Different Questions, Different Answers? Media Use and Media Credibility." *Journalism Quarterly,* 64: 28–44. DOI:10.1177/107769908706400104.

Schaeffer, Nora Cate. 1991. "Hardly Ever or Constantly: Group Comparisons Using Vague Quantifiers." *Public Opinion Quarterly,* 55: 395–423. DOI:10.1086/269270.

Schwarz, Norbert, Carla E. Grayson, and Barbel Knauper. 1998. "Formal Features of Rating Scales and the Interpretation of Question Meaning." *International Journal of Public Opinion Research,* 10: 177–183. DOI:10.1093/ijpor/10.2.177.

Sigal, Leon V. 1973. *Reporters and Officials.* Lexington, MA: D. C. Heath & Co.

Sobel, Michael E. 1997. "Modeling Symmetry, Asymmetry, and Change in Ordered Scales with Midpoints Using Adjacent Category Logit Models for Discrete Data." *Sociological Methods & Research,* 26: 213–232. DOI:10.1177/0049124197026002004.

Stempel, Carl, Thomas Hargrove, and Guido H. Stempel III. 2007. "Media Use, Social Structure, and Belief in 9/11 Conspiracy Theories." *Journalism & Mass Communication Quarterly,* 84: 353–372. DOI:10.1177/107769900708400210.

Stokes, Maura E., Charles S. Davis, and Gary G. Koch. 2012. *Categorical Data Analysis Using SAS.* 3rd ed. Cary, NC: SAS Institute.

Strom, Renee E., and Franklin J. Boster. 2011. "Dropping Out of High School: Assessing the Relationship between Supportive Messages from Family and Education Attainment." *Communication Reports*, 24: 25–37. DOI:10.1080/08934215.2011.554623.

The American National Election Studies. 2008. American National Election Study: ANES Pre- and Post-Election Survey. ICPSR25383-v2. Ann Arbor, MI: Inter-university Consortium for Political and Social Research [distributor], 2012–08–30. DOI:10.3886/ICPSR25383.v2.

Wanke, Michaela. 2002. "Conversational Norms and the Interpretation of Vague Quantifiers." *Applied Cognitive Psychology*, 16: 301–307. DOI:10.1002/acp.787.

Wilson, Thomas P. 1971. "Critique of Ordinal Variables." *Social Forces*, 49: 432–444. DOI:10.1093/sf/49.3.432.

Winship, Christopher, and Robert D. Mare. 1984. "Regression Models with Ordinal Variables." *American Sociological Review*, 49: 512–525.

Wright, Daniel B., George D. Gaskell, and Colm A. O'Muircheartaigh. 1994. "How Much is 'Quite a Bit'?: Mapping Between Numerical Values and Vague Quantifiers." *Applied Cognitive Psychology*, 8: 479–496. DOI:10.1002/acp.2350080506.

9

Probit Analysis

Probit analysis originated in studies of biological assay (Bliss 1934), which Finney (1952, 1) defined as "the measurement of the potency of any stimulus, physical, chemical or biological, physiological or psychological, by means of the reactions which it produces in living matter." As an example, a researcher might use the probit model in a study of toxicology, identifying the point at which a pesticide eliminates a certain type of insect (Agresti 1990, 102). In such a study, the researcher would increase doses of the pesticide until the desired effect – elimination of the pest – occurs. Like the binary logistic regression model, which uses the logit, or log of the odds, to transform a dependent variable into a linear function of explanatory measures (see Allison 2015), probit analysis uses the inverse of the cumulative normal distribution as a link function (see Aldrich and Nelson 1984, Bishop, Fienberg, and Holland 1975, 367, Stokes, Davis, and Koch 2012, 345–346). As Azen and Walker (2011, 121–124) explained, the probit model assumes a continuous scale underlying a binary response measure, and the assumption of a latent measure can prove useful when a variable lacks a conceptual dichotomy. Attitudes toward abortion, for example, may be somewhat nuanced (e.g., affected by occurrences of incest or rape), and accordingly a statistical model should assume the potential for attitudinal distinctions while treating a variable as dichotomous. By using the inverse of the cumulative normal distribution, a probit model identifies a threshold at which attitudes shift (see Hanushek and Jackson 1977, 204, Long 1997, 40).[1]

As a generalized linear model (McCullagh and Nelder 1989), probit analysis contains ordinal (Daykin and Moffatt 2002, McKelvey and Zavoina 1975, Winship and Mare 1984) and multinomial (Dow and Endersby 2004) extensions. Computationally, these techniques are more complex than logit-based models, and indeed this relative complexity helps to explain why probit analyses appear

Categorical Statistics for Communication Research, First Edition. Bryan E. Denham.
© 2017 John Wiley & Sons, Inc. Published 2017 by John Wiley & Sons, Inc.
Companion Website: www.wiley.com/go/denham/categorical_statistics

less frequently in social science research. The multinomial probit model is especially complex and is not discussed at length in the current chapter. In contrast, the binary and ordinal probit models have proven useful in certain types of communication studies and, as Daykin and Moffatt (2002) noted, the procedures perform especially well with survey data (see also Borooah 2002, Liao 1994). Tang, He, and Tu (2012, 157) added that probit models continue to be popular in analyses involving latent variables; the authors cited mixed-effects models for longitudinal and clustered data as examples. The current chapter reviews published studies in communication and discusses the fundamentals of probit analysis with data from the 2008 American National Election Study. The chapter includes SPSS instruction for binary and ordinal probit analyses as part of this discussion.

Examples of Published Research

Scholars have used binary and polytomous probit models in studies of political and health communication, risk and strategic communication, as well as media economics and policy. In political communication, Freedman and Goldstein (1999) examined the impact of negative advertising on the probability of voting. Jaeger (2008) studied whether a left–right political orientation predicted attitudes toward a welfare state, while Melgar, Rossi, and Smith (2010) used probit analysis in a study addressing perceptions of corruption. Mutz and Martin (2001) analyzed news media as a source of diverse political perspectives, while Fullerton, Dixon, and Borch (2007) modeled determinants of vote overreporting (i.e., false claims of voting among survey respondents). Miller and Krosnick (2000) studied factors associated with presidential evaluation, while Gunther (1992) determined that group membership plays an important role in assessments of how news media cover social groups.

In health communication, Babalola and Kincaid (2009) proposed a modified probit model for the evaluation of health communication programs, and Hutchinson et al. (2006) studied cost-effectiveness in the Smiling Sun media campaign in rural Bangladesh. In a risk analysis, Smith and Desvousges (1990) analyzed whether individuals would pay for a licensed technician to examine potential radon problems in the home. Studying strategic communication, Sommerfeldt (2013) analyzed relationships between resource mobilization and activism strategies, while Vigderhous (1977) used a probit model to estimate the point at which individuals become aware of a product. Wirth and Bloch (1989) used the technique in a study addressing why groups of individuals purchase cable television, and Fu (2003) analyzed why media companies opt to publish (or not publish) a Sunday newspaper. Finally, Tillema, Dijst, and Schwanen (2010) used probit analysis in studying face-to-face and electronic communications.

Probit Analysis: Fundamentals

Much of this text has focused on statistical techniques that transform an outcome measure using the logit, or log of the odds. In most instances, logit models function very well, and statisticians such as Haberman (1978, 314) have questioned the need to depart from logistic regression techniques when working with categorical response measures. But probit analysis can prove useful when a categorical outcome measure is thought to contain an underlying continuous scale and a normal distribution. In survey research, questions about abortion, capital punishment, and gun control often contain binary response options, yet attitudes may vary considerably. By assuming the presence of a latent continuous scale (see DeMaris 2004, 251–260, Powers and Xie 2000, 41–85), the probit model links substantive concerns with statistical estimation (McKelvey and Zavoina 1975).

As indicated in Chapter 4, generalized linear models consist of three components: random and systematic components as well as a link function (McCullagh and Nelder 1989). As one type of generalized linear model, probit analysis assumes a binomial distribution as its random component, one or more categorical or continuous measures as the systematic component(s), and the inverse of the cumulative normal distribution (i.e., probit) as a link function. Like the logit model, probit analysis contains estimates based on maximum likelihood, but unlike the logit model, probit assumes an underlying quantitative measure associated with the normal distribution. By assuming a qualitative variable, logit allows odds ratios to summarize associations, whereas a probit model reveals the extent to which a one-unit change in an explanatory variable affects the level of a response measure, given a normal distribution. In general, probit estimates do not contain the same magnitude as those produced in logit analyses, as Agresti (2007, 73) explained:

> This is because their link functions transform probabilities to scores from standard versions of the normal and logistic distribution, but those two distributions have different spread. The standard normal distribution has a mean of 0 and standard deviation of 1. The standard logistic distribution has a mean of 0 and standard deviation of 1.8. When both models fit well, parameter estimates in logistic regression models are approximately 1.8 times those in probit models.

Statistically, Aldrich and Nelson (1984, 48) explained, the binary probit model focuses on the value of parameter P, or the probability that $Y=1$, which is expressed as $P = P(Y=1)$. They further explained that Y depends on K observable variables X_k, $k=1,\ldots, K$. The assumption that explanatory variables account for variance in P can be expressed as $P = P(Y=1|X_1,\ldots, X_k)$, or $P(Y|X)$, with X representing the set of K explanatory measures. The probit model is then expressed as:

$$P(Y\,|\,X) = \Phi\left(X'\beta\right)$$

where Φ represents the cumulative distribution function of the standard normal distribution, and β indicates model parameters to be estimated with maximum likelihood. A key point for the current chapter is that a probit link transforms probabilities into z-scores, which are based on the standard normal distribution. As Norusis (2005, 92) explained, "If half the subjects respond at a particular [level of a stimulus], the corresponding probit value is 0, since half of the area in a standard normal curve falls below a z score of 0. If the observed proportion is 0.95, the corresponding probit value is 1.64."

Regarding goodness of fit, a probit analysis produces a likelihood ratio statistic, with a chi-square value indicating whether a model containing the explanatory measure(s) fits better than a model containing only the intercept (Aldrich and Nelson 1984, 55). Probit analysis also informs researchers about the relative contribution of each explanatory measure. The following section addresses the binary probit model using two techniques in SPSS.

Binary Probit Analysis

The binary probit model analyzes the effects of categorical and continuous explanatory measures on a dichotomous response variable. In statistical packages such as SPSS, the model can be estimated through Generalized Linear Models or through one or more Regression procedures.[2] This chapter uses data from the 2008 American National Election Study (The American National Election Studies 2008) to demonstrate binary probit analysis in SPSS, utilizing the Generalized Linear Models and PLUM procedures, respectively, and also covers ordinal probit analysis using the PLUM model.[3] In the binary analyses, probit models contain a dependent measure indicating whether individuals had discussed politics with friends and family, with independent measures including sex, race, and attitudes toward the United States (i.e., Optimistic, Pessimistic, Neither Optimistic nor Pessimistic). For the ordinal probit model, a variable indicating the extent to which individuals trust media to report issues fairly (i.e., Almost Always, Most of the Time, Some of the Time, Almost Never) functioned as a response measure, with the same explanatory variables.

Figure 9.1 contains a screenshot for Generalized Linear Models in SPSS (Analyze > Generalized Linear Models > Generalized Linear Models). As shown in the figure, multiple procedures appear in the Type of Model window, including Binary probit. In Chapter 10, which focuses on Poisson regression, options under Counts (i.e., Poisson loglinear and Negative binomial with log link) will be utilized. One can also fit Binary logistic, ordinal and scale, and mixed and custom models.

In addition to Type of Model, tabs in the Generalized Linear Models procedure include Response, Predictors, Model, Estimation, Statistics, EM

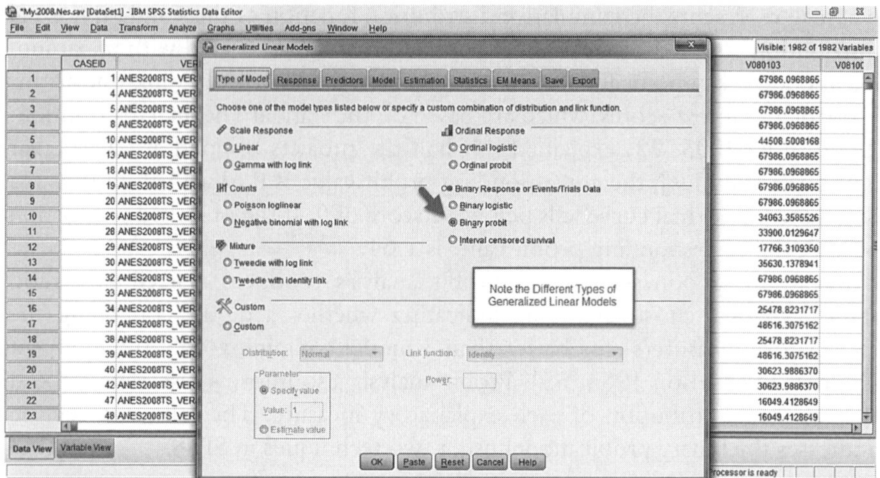

Figure 9.1 SPSS screenshot for Generalized Linear Models. Source: SPSS® Reprints Courtesy of International Business Machines Corporation, © 2014 International Business Machines Corporation

Means, Save, and Export. A probit analysis requires the researcher to specify one response measure and one or more predictor measures in the respective windows.

After specifying variables, one can open the Model tab, shown in the window that appears in Figure 9.2. Here the researcher can choose to model Main effects, as in the current example, or select more advanced models containing variable interactions. The Estimation window allows the researcher to specify criteria for parameter estimation, iterations, and convergence. In most cases, default options function in a satisfactory manner, although researchers can make adjustments if necessary. The Statistics window allows the researcher to specify information to be included in the output, including confidence intervals, descriptive statistics, parameter estimates, and so forth. EM Means displays estimated marginal means for factors and factor interactions, while Save and Export allow the researcher to specify additional information to be included in analyses and statistical output.

Table 9.1 includes output for the current example. In this table, Model Information confirms the dependent variable and the level of the dependent variable modeled as well as the binomial distribution and the probit link function. Then comes a series of goodness-of-fit statistics, and researchers should pay close attention to the Deviance and Pearson Chi-Square values. In general, a suitable model will contain a Value/df ratio close to 1.0, and the current example offers a reasonably good fit given three explanatory variables. The Omnibus Test indicates that a model containing the three explanatory measures

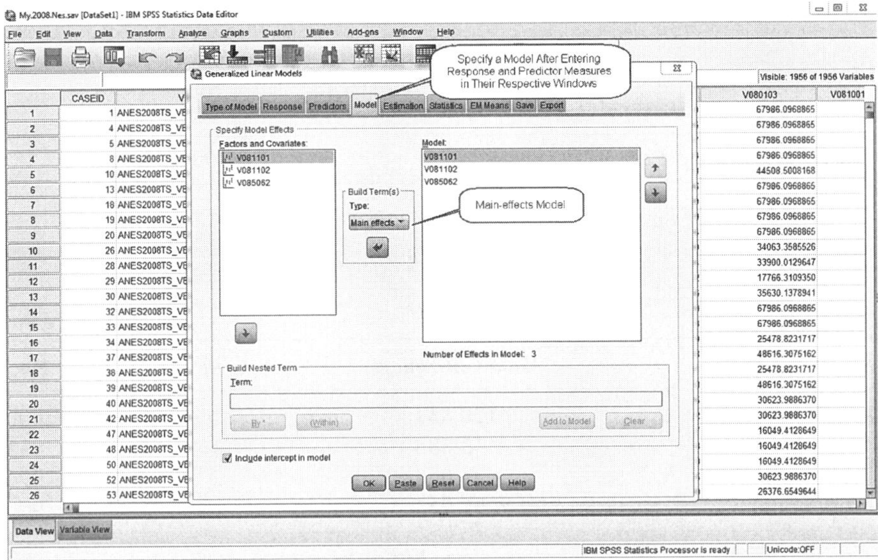

Figure 9.2   SPSS screenshot for Model design in Generalized Linear Models. Source: SPSS® Reprints Courtesy of International Business Machines Corporation, © 2014 International Business Machines Corporation

Table 9.1   SPSS output for binary probit model in Generalized Linear Models procedure

Model Information

Dependent Variable	Ever discuss politics with family or friends[a]
Probability Distribution	Binomial
Link Function	Probit

[a] The procedure models Yes as the response, treating No as the reference category.

Goodness of Fit[b]

	Value	df	Value/df
Deviance	10.928	12	.911
Scaled Deviance	10.928	12	
Pearson Chi-Square	10.883	12	.907
Scaled Pearson Chi-Square	10.883	12	
Log Likelihood[c]	−38.606		
Akaike's Information Criterion (AIC)	89.212		
Finite Sample Corrected AIC (AICC)	89.294		
Bayesian Information Criterion (BIC)	118.835		
Consistent AIC (CAIC)	124.835		

[b] Information criteria are in small-is-better form.
[c] The full log likelihood function is displayed and used in computing information criteria.

(Continued)

Table 9.1   (Continued)

	Omnibus Test[a]		
Likelihood Ratio Chi-Square		*df*	*Sig.*
63.129		5	.000

[a]Compares the fitted model against the intercept-only model.

Test of Model Effects

	Type III		
Source	*Wald Chi-Square*	*df*	*Sig.*
Intercept	120.851	1	.000
Sex	3.785	1	.052
Race	.008	2	.996
Attitude about US	57.925	2	.000

Parameter Estimates

Parameter	*B*	*SE*	*95% Wald Confidence Interval*		*Hypothesis Test*		
			Lower	*Upper*	*Wald Chi-Square*	*df*	*Sig.*
Intercept	.151	.1395	−.122	.425	1.178	1	.278
Male	.173	.0889	−.001	.347	3.785	1	.052
White	−.006	.1372	−.275	.262	.002	1	.963
Black	.002	.1514	−.294	.299	.000	1	.987
Optimistic	.719	.0958	.531	.907	56.280	1	.000
Pessimistic	.576	.1368	.308	.844	17.705	1	.000

fits better than a model containing only the intercept, and the Test of Model Effects offers a glimpse of whether each predictor contributes in a statistically significant manner. In this case, the positive parameter estimate for males indicates they appeared more likely to respond "yes" to the question constituting the dependent variable; however, as indicated in Parameter Estimates, the difference between males and females was not significant at $p < .05$. Additionally, differences did not emerge across race categories, but the significant positive estimates for Optimistic and Pessimistic indicate that individuals in these two categories were more likely to discuss politics than individuals who stated they were Neither Optimistic nor Pessimistic about the United States. In addition to parameter estimates, the last section includes standard errors, confidence intervals, and the results of Wald chi-square tests.

To inform the binary probit model estimated through the SPSS Generalized Linear Models procedure, Table 9.2 contains the same model estimated with SPSS PLUM.[4] Beginning with Model Fitting Information, one might notice the Chi-Square value under Likelihood Ratio Tests; as in the Omnibus Test from Table 9.1, the Chi-Square value is 63.129 with 5 degrees of freedom and a *p*-value of less than .001. This output confirms the model containing three explanatory variables fits better than a model containing only the intercept. Next, the Chi-Square values

Table 9.2 SPSS output for binary probit model in PLUM procedure

Model Fitting Information

Model	Model Fitting Criteria −2 Log Likelihood	Likelihood Ratio Tests Chi-Square	df	Sig.
Intercept Only	140.341			
Final	77.212	63.129	5	.000

Goodness-of-Fit

	Chi-Square	df	Sig.
Pearson	10.883	12	.539
Deviance	10.928	12	.535

Pseudo R-Square

Cox and Snell	.059
Nagelkerke	.088
McFadden	.055

Parameter Estimates

	B	SE	Wald	df	Sig.	95% Confidence Interval Lower	Upper
Threshold							
Discussed Politics	.151	.140	1.174	1	.279	−.122	.425
Location							
Male	−.173	.089	3.779	1	.052	−.347	.001
White	.006	.137	.002	1	.963	−.263	.275
Black	−.002	.151	.000	1	.987	−.299	.294
Optimistic	−.719	.096	56.308	1	.000	−.907	−.531
Pessimistic	−.576	.137	17.743	1	.000	−.843	−.308

for Pearson and Deviance agree with the statistics presented in Table 9.1, and also included in Table 9.2 are Pseudo R-Square statistics indicating an amount of variance explained by three predictor variables (see Hagle and Mitchell 1992).

Lastly, one observes in Table 9.2 parameter estimates reflecting those shown in Table 9.1. It is important to recognize the difference in positive and negative estimates, which was to be expected given the manner in which PLUM reports predictor effects on a single response measure. The "order" of the dependent variable was yes and no, and the negative estimate for males shows their greater likelihood of indicating the initial value. The same applies to other variables and their categories. Additional statistics include standard errors, Wald chi-square tests, and 95% confidence intervals, consistent with the PLUM-based ordinal logistic regression analyses presented in Chapter 8.

Ordinal Probit Analysis

Ordinary least squares (OLS) regression assumes equal intervals between data points as well as homogeneity of variance and a normal distribution of error terms. In the social sciences, meeting each of these assumptions can prove challenging, especially when studies treat ordinal response measures as continuous, quantitative variables. Like the ordinal logistic regression model, ordinal probit analysis solves some of the problems associated with OLS modeling. Probit analysis accommodates variation in the meaning respondents assign to vague quantifiers as well as potential differences in the size of intervals between data points (Daykin and Moffatt 2002, 159). Additionally, unlike OLS models, ordinal probit analyses do not model survey responses directly, and therefore the wording of a question is less likely to influence statistical results. As Daykin and Moffatt (2002, 159) suggested, "The distribution over the population of the underlying attitude, which is the focus of analysis, should be invariant to the wording of the question. Because the ordered probit model estimates the parameters of the underlying distribution, rather than the response itself, any such framing effects are likely to be avoided."

Discussing the ordinal probit model, Hanushek and Jackson (1977, 211) explained how ordered categories result in additional thresholds, or breaks in the data. Regarding notation, the authors stated that when a dependent variable contains three levels, a model assumes two thresholds, U_t^1 and U_t^2. With units assigned to one of three groups (i.e., 1, 2, 3) and asterisks indicating standardized scores, Hanushek and Jackson (1977, 211) specified the classification of Y_t as:

$$Y_t = \begin{cases} 3 : U_t^{2*} \leq Z_t = X_t\beta^* \\ 2 : U_t^{1*} \leq Z_t = X_t\beta^* < U_t^{2*} \\ 1 : Z_t = X_t\beta^* < U_t^{1*} \end{cases}$$

As Hanushek and Jackson (1977, 212) summarized, "The model, then, considers all categories of the dependent variable to be aggregations of a range of the underlying variable, meaning they constitute *ordered* categories."

Table 9.3 contains output for an ordinal probit analysis conducted with the PLUM procedure in SPSS. First, in Model Fitting Information, the –2 Log Likelihood values and Likelihood Ratio Test indicate that a model testing the effects of sex, race, and attitudes toward the United States on trust in media fits better than a model containing only the intercept. Pearson and Deviance statistics confirm the fit of this model, and three Pseudo R-Square measures show an approximate amount of variance explained. Moving to Parameter Estimates, the last category for each variable serves as a reference for others, and in this example, the positive estimates for Males, Whites, and Pessimistic respondents indicate their tendency to move in the same direction as the dependent variable, meaning individuals in these categories were less trusting of the media than were respondents in other categories. Additional statistics include standard errors, Wald chi-square statistics, and 95% confidence intervals.

In addition to the statistics reported in Table 9.3, SPSS produces observed and expected cell counts as well as residual values. Examining the values shown Tables 9.4a and 9.4b, the three-variable probit model fit the data reasonably well, with relatively small differences between observed and expected counts.

Table 9.3   SPSS output for ordinal probit regression model

| | ***Model Fitting Information*** | | | |

	Model Fitting Criteria	Likelihood Ratio Tests		
Model	*–2 Log Likelihood*	*Chi-Square*	*df*	*Sig.*
Intercept Only	324.801			
Final	247.592	77.110	5	.000

Goodness-of-Fit			
	Chi-Square	*df*	*Sig.*
Pearson	49.416	46	.338
Deviance	47.190	46	.424

Pseudo R-Square	
Cox and Snell	.037
Nagelkerke	.042
McFadden	.018

(*Continued*)

Table 9.3 (Continued)

| | | | *Parameter Estimates* | | | | | |
| | | | | | | | 95% Confidence Interval | |
	B	*SE*	*Wald*	*df*	*Sig.*	*Lower*	*Upper*
Threshold							
Trust Media Just About Always	−1.586	.090	310.012	1	.000	−1.763	−1.410
Trust Media Most of the Time	−.228	.083	7.634	1	.006	−.390	−.066
Trust Media Some of the Time	1.484	.088	283.765	1	.000	1.311	1.656
Location							
Male	.131	.049	7.093	1	.008	.035	.227
White	.226	.076	8.740	1	.003	.076	.375
Black	−.045	.085	.278	1	.598	−.211	.121
Optimistic	−.260	.056	21.454	1	.000	−.370	−.150
Pessimistic	.102	.078	1.730	1	.188	−.050	.255

As indicated in previous chapters, reports of cell frequencies allow researchers to confirm the directionalities of parameter estimates, and residuals indicate where marked differences occur between observed and expected frequencies. The tables also allow researchers to consider the general dispersion of responses and whether categories have sufficient observations for analyses.

Multinomial Probit Analysis

Although the current chapter does not cover the multinomial probit model in detail, readers interested in this approach might begin by reviewing the work of Hausman and Wise (1978) as well as studies in transportation (Horowitz 1991), economics (Geweke, Keane, and Runkle 1994) and political science (Alvarez and Nagler 1995, Dow and Endersby 2004). Powers and Xie (2000, 248) noted that LIMDEP 7.0 has proven useful in estimating multinomial probit models, but for purposes of communication research, multinomial logistic regression, covered in Chapter 7, generally proves satisfactory for most analyses containing multi-level nominal variables.

Interactions

Like logistic regression models, probit analysis will test user-specified interactions among explanatory variables (see Ai and Norton 2003, Brambor, Clark, and Golder 2006). As Huang and Shields (2000, 88) argued, "Interactions

Table 9.4a SPSS frequency output for ordinal probit regression model

Sex	Race	Attitude about US		How Often Trust Media to Report News Fairly			
				Almost Always	Most of Time	Some of Time	Almost Never
Male	White	Optimistic	Observed	14	111	153	26
			Expected	14.056	99.287	165.533	25.125
			Residual	−.015	1.433	−1.443	.182
		Pessimistic	Observed	2	25	65	23
			Expected	2.350	25.953	69.140	17.557
			Residual	−.230	−.213	−.789	1.411
		Neither	Observed	6	34	90	20
			Expected	3.904	38.011	88.614	19.471
			Residual	1.075	−.753	.230	.129
	Black	Optimistic	Observed	9	55	65	4
			Expected	10.499	53.132	62.894	6.475
			Residual	−.482	.331	.366	−.997
		Pessimistic	Observed	2	3	8	1
			Expected	.532	4.208	7.894	1.366
			Residual	2.053	−.704	.057	−.330
		Neither	Observed	4	21	31	7
			Expected	2.975	20.754	34.160	5.111
			Residual	.609	.066	−.799	.872
	Other Race	Optimistic	Observed	3	21	30	4
			Expected	4.209	22.508	28.187	3.096
			Residual	−.612	−.406	.476	.528
		Pessimistic	Observed	0	3	9	3
			Expected	.516	4.318	8.582	1.583
			Residual	−.731	−.752	.218	1.191
		Neither	Observed	3	12	22	3
			Expected	1.719	12.673	22.086	3.522
			Residual	.999	−.229	−.027	−.291

Table 9.4b SPSS frequency output for ordinal probit regression model

Sex	Race	Attitude about US		How Often Trust Media to Report News Fairly			
				Almost Always	Most of Time	Some of Time	Almost Never
Female	White	Optimistic	Observed	17	148	206	21
			Expected	23.665	142.258	200.805	25.272
			Residual	−1.413	.603	.525	−.878
		Pessimistic	Observed	8	24	96	13
			Expected	3.920	36.846	82.769	17.465
			Residual	2.090	−2.462	2.263	−1.141
		Neither	Observed	6	56	116	18
			Expected	6.861	56.843	111.878	20.418
			Residual	−.335	−.133	.595	−.565
	Black	Optimistic	Observed	15	79	88	7
			Expected	18.904	81.375	81.758	6.963
			Residual	−.946	−.349	.916	.014
		Pessimistic	Observed	3	8	10	1
			Expected	1.102	7.424	11.782	1.692
			Residual	1.855	.260	−.762	−.554
		Neither	Observed	9	34	42	10
			Expected	5.850	34.735	48.410	6.005
			Residual	1.344	−.157	−1.315	1.684
	Other Race	Optimistic	Observed	8	39	30	4
			Expected	7.484	34.050	36.178	3.288
			Residual	.198	1.114	−1.381	.401
		Pessimistic	Observed	1	7	6	0
			Expected	.639	4.548	7.642	1.170
			Residual	.462	1.399	−.882	−1.130
		Neither	Observed	1	16	22	4
			Expected	2.422	15.197	22.416	2.965
			Residual	−.941	.256	−.127	.623

assumed by an S-shaped curve are not an evil to avoid but an approximation of the underlying data-generation process of binary responses." Interactions can also be tested in the ordinal probit model, and their inclusion should follow steps taken in ordinal logistic regression analyses (i.e., steps taken in the PLUM procedure). In the present chapter, interactive effects were tested but none showed statistical significance.

Chapter Summary

This chapter has focused on binary and ordinal probit models, which assume an underlying continuous variable and the inverse of the cumulative normal distribution as a link function. In SPSS, the binary probit model can be fit in the Generalized Linear Models procedure and can also be tested through PLUM. The binary probit model proves useful when dichotomous dependent variables are thought to contain a degree of internal variation, as in the case of support/ lack of support for social legislation. The ordinal probit model, fit through PLUM, is similarly flexible. Models containing multi-level nominal response measures should be tested with multinomial logistic regression analyses.

Chapter Exercises

1. In Chapter 6, the first exercise question focused on time period, presence on the front page, and dateline as determinants of anonymous attribution in news reports. For the current question, use the data from that item in Chapter 6 to test a binary probit model for the same effects. Use the PLUM procedure in SPSS and report −2 log likelihood values for the Intercept Only and Final models, as well as the Pearson goodness-of-fit test and three R-squared estimates. Also report parameter estimates and their standard errors, in addition to Wald values, significance tests, and 95% confidence intervals. Are the probit and logit models similar, as one would anticipate?

2. Chapter 8 included an exercise based on data gathered in a CBS News poll (CBS News 2008), with a three-level ordinal response measure indicating whether members of racial minorities were Very Interested, Somewhat Interested, or Not at All Interested in professional baseball. The current exercise contains sex and race as explanatory measures, as opposed to sex and education level. Use the SPSS PLUM procedure to fit an ordinal probit model to the data below, making certain to test the interaction between sex and race. Then report −2 log likelihood values for the Intercept Only and Final models, as well as the Pearson goodness-of-fit test and three R-squared estimates.

212   *Categorical Statistics for Communication Research*

Also report parameter estimates and their standard errors, in addition to Wald values, significance tests, and 95% confidence intervals. What do the data reveal about sex and race as predictors of interest in baseball?

ID	B S R	ID	B S R	ID	B S R	ID	B S R	ID	B S R	ID	B S R	ID	B S R
001	1 1 1	026	2 1 1	051	2 2 1	076	2 2 2	101	3 1 1	126	3 2 1	151	3 2 1
002	1 1 1	027	2 1 1	052	2 2 1	077	2 2 2	102	3 1 2	127	3 2 1	152	3 2 1
003	1 1 1	028	2 1 1	053	2 2 1	078	3 1 1	103	3 1 2	128	3 2 1	153	3 2 1
004	1 1 1	029	2 1 1	054	2 2 1	079	3 1 1	104	3 1 2	129	3 2 1	154	3 2 1
005	1 1 1	030	2 1 1	055	2 2 1	080	3 1 1	105	3 1 2	130	3 2 1	155	3 2 1
006	1 1 1	031	2 1 1	056	2 2 1	081	3 1 1	106	3 1 2	131	3 2 1	156	3 2 1
007	1 1 1	032	2 1 1	057	2 2 1	082	3 1 1	107	3 1 2	132	3 2 1	157	3 2 1
008	1 1 1	033	2 1 1	058	2 2 1	083	3 1 1	108	3 1 2	133	3 2 1	158	3 2 2
009	1 1 1	034	2 1 1	059	2 2 1	084	3 1 1	109	3 1 2	134	3 2 1	159	3 2 2
010	1 1 2	035	2 1 1	060	2 2 1	085	3 1 1	110	3 1 2	135	3 2 1	160	3 2 2
011	1 1 2	036	2 1 1	061	2 2 1	086	3 1 1	111	3 2 1	136	3 2 1	161	3 2 2
012	1 2 1	037	2 1 1	062	2 2 1	087	3 1 1	112	3 2 1	137	3 2 1	162	3 2 2
013	1 2 1	038	2 1 1	063	2 2 1	088	3 1 1	113	3 2 1	138	3 2 1	163	3 2 2
014	1 2 1	039	2 1 1	064	2 2 1	089	3 1 1	114	3 2 1	139	3 2 1	164	3 2 2
015	1 2 1	040	2 1 1	065	2 2 1	090	3 1 1	115	3 2 1	140	3 2 1	165	3 2 2
016	1 2 1	041	2 1 2	066	2 2 1	091	3 1 1	116	3 2 1	141	3 2 1	166	3 2 2
017	1 2 1	042	2 1 2	067	2 2 1	092	3 1 1	117	3 2 1	142	3 2 1	167	3 2 2
018	1 2 2	043	2 1 2	068	2 2 1	093	3 1 1	118	3 2 1	143	3 2 1	168	3 2 2
019	1 2 2	044	2 1 2	069	2 2 1	094	3 1 1	119	3 2 1	144	3 2 1	169	3 2 2
020	1 2 2	045	2 1 2	070	2 2 1	095	3 1 1	120	3 2 1	145	3 2 1	170	3 2 2
021	1 2 2	046	2 1 2	071	2 2 2	096	3 1 1	121	3 2 1	146	3 2 1	171	3 2 2
022	1 2 2	047	2 1 2	072	2 2 2	097	3 1 1	122	3 2 1	147	3 2 1	172	3 2 2
023	1 2 2	048	2 1 2	073	2 2 2	098	3 1 1	123	3 2 1	148	3 2 1	173	3 2 2
024	1 2 2	049	2 1 2	074	2 2 2	099	3 1 1	124	3 2 1	149	3 2 1	174	3 2 2
025	2 1 1	050	2 1 2	075	2 2 2	100	3 1 1	125	3 2 1	150	3 2 1	175	3 2 2

Note: Excel file containing data available on companion website.

Category codes:
Interest in Baseball (B):   1 = Very Interested, 2 = Somewhat Interested, 3 = Not at all Interested
Sex (S):   1 = Males, 2 = Females
Racial Minority (R):   1 = African Americans, 2 = Members of Other Racial Minority

Notes

1   As Finney (1952, 8) explained, "For any one subject, under controlled conditions, there will be a certain level of intensity below which the response does not occur and above which the response occurs; in psychology such a value is designated the *threshold* or *limen*, but in pharmacology and toxicology the term *tolerance* seems more appropriate."

2   SPSS offers a Probit model among its Regression choices, but the technique is designed specifically for grouped-response dose analyses (e.g., toxicology studies identifying the level at which a chemical substance proves fatal). See Norusis (2005, 91–101) for instruction.
3   Although the PLUM model is used primarily for ordinal logistic regression, as well as ordinal probit regression, it will also fit a binary probit model.
4   In the SPSS PLUM procedure, one can click on the Options tab and switch the analysis from Logit to Probit. Additional models such as Complementary log-log are also available.

References

Agresti, Alan. 1990. *Categorical Data Analysis*. New York: John Wiley & Sons.

Agresti, Alan. 2007. *Introduction to Categorical Data Analysis*, 2nd ed. Hoboken, NJ: John Wiley & Sons.

Ai, Chunrong, and Edward Norton. 2003. "Interaction Terms in Logit and Probit Models." *Economics Letters*, 80: 123–129. DOI:10.1016/S0165-1765(03)00032-6.

Aldrich, John H., and Forrest D. Nelson. 1984. *Linear Probability, Logit, and Probit Models*. Newbury Park, CA: Sage.

Allison, Paul. 2015. "What's So Special about Logit?" *Statistical Horizons*. Retrieved from http://statisticalhorizons.com/whats-so-special-about-logit.

Alvarez, R. Michael, and Jonathan Nagler. 1995. "Issues, Economics, and the Perot Candidacy: Voter Choice in the 1992 Presidential Election." *American Journal of Political Science*, 39: 714–744.

Azen, Razia, and Cindy M. Walker. 2011. *Categorical Data Analysis for the Behavioral and Social Sciences*. New York: Routledge.

Babalola, Stella, and D. Lawrence Kincaid. 2009. "New Methods for Estimating the Impact of Health Communication Programs." *Communication Methods and Measures*, 3: 61–83. DOI:10.1080/19312450902809706.

Bishop, Yvonne M. M., Stephen E. Fienberg, and Paul W. Holland. 1975. *Discrete Multivariate Analysis: Theory and Practice*. Cambridge, MA: MIT Press.

Bliss, C. I. 1934. "The Method of Probits." *Science*, 79: 38–39. DOI:10.1126/science.79.2037.38.

Borooah, Vani K. 2002. *Logit and Probit: Ordered and Multinomial Models*. Thousand Oaks, CA: Sage.

Brambor, Thomas, William Roberts Clark, and Matt Golder. 2006. "Understanding Interaction Models: Improving Empirical Analyses." *Political Analysis*, 14: 63–82. DOI:10.1093/pan/mpi014.

CBS News. 2008. CBS News Monthly Poll #1, March 2008. ICPSR26144-v1. Ann Arbor, MI: Inter-university Consortium for Political and Social Research [distributor], 2009-10-08. DOI:10.3886/ICPSR26144.v1.

Daykin, Anne R., and Peter G. Moffatt. 2002. "Analyzing Ordered Responses: A Review of the Ordered Probit Model." *Understanding Statistics*, 1: 157–166. DOI:10.1207/S15328031US0103_02.

DeMaris, Alfred. 2004. *Regression with Social Data: Modeling Continuous and Limited Response Variables*. Hoboken, NJ: John Wiley & Sons.

Dow, Jay K., and James W. Endersby. 2004. "Multinomial Probit and Multinomial Logit: A Comparison of Choice Models for Voting Research." *Electoral Studies*, 23: 107–122. DOI:10.1016/S0261–3794(03)00040–4.

Finney, D. J. 1952. *Probit Analysis*, 2nd ed. Cambridge: Cambridge University Press.

Freedman, Paul, and Ken Goldstein. 1999. "Measuring Media Exposure and the Effects of Negative Campaign Ads." *American Journal of Political Science*, 43: 1189–1208.

Fu, W. Wayne. 2003. "Deliver on Sunday? Analysis of Daily Newspapers' Entry to Sunday Edition Publishing." *Journal of Media Economics*, 16(3): 189–205. DOI:10.1207/S15327736ME1603_4.

Fullerton, Andrew S., Jeffrey C. Dixon, and Casey Borch. 2007. "Bringing Registration into Models of Vote Overreporting." *Public Opinion Quarterly*, 71: 649–660. DOI:10.1093/poq/nfm040.

Geweke, John, Michael Keane, and David Runkle. 1994. "Alternative Computation Approaches to Inference in the Multinomial Probit Model." *Review of Economics and Statistics*, 76: 609–632.

Gunther, Albert C. 1992. "Biased Press or Biased Public? Attitudes toward Media Coverage of Social Groups." *Public Opinion Quarterly*, 56: 147–167. DOI:10.1086/269308.

Haberman, Shelby J. 1978. *Analysis of Qualitative Data*. New York: Academic Press.

Hagle, Timothy M., and Glenn E. Mitchell. 1992. "Goodness-of-Fit Measures for Probit and Logit." *American Journal of Political Science*, 36: 762–784.

Hanushek, Eric A., and John E. Jackson. 1977. *Statistical Methods for Social Scientists*. New York: Academic Press.

Hausman, Jerry A., and David A. Wise. 1978. "A Conditional Probit Model for Qualitative Choice: Discrete Decisions Recognizing Interdependence and Heterogeneous Preferences." *Econometrica*, 46: 403–426.

Horowitz, Joel L. 1991. "Reconsidering the Multinomial Probit Model." *Transportation Research*, B 25: 433–438. DOI:10.1016/0191–2615(91)90036-I.

Huang, Chi, and Todd G. Shields. 2000. "Interpretation of Interaction Effects in Logit and Probit Analyses: Reconsidering the Relationship Between Registration Laws, Education, and Voter Turnout." *American Politics Quarterly*, 28: 80–95. DOI:10.1177/1532673X00028001005.

Hutchinson, Paul, Peter Lance, David K. Guilkey, Mohammad Shahjahan, and Shahida Haque. 2006. "Measuring the Cost-Effectiveness of a National Health Communication Program in Rural Bangladesh." *Journal of Health Communication*, 11: 91–121. DOI:10.1080/10810730600974647.

Jaeger, Mads Meier. 2008. "Does Left-Right Orientation Have a Causal Effect on Support for Redistribution? Causal Analysis with Cross-Sectional Data Using Instrumental Variables." *International Journal of Public Opinion Research*, 20: 363–374. DOI:10.1093/ijpor/edn030.

Liao, Tim Futing. 1994. *Interpreting Probability Models: Logit, Probit, and Other Generalized Linear Models*. Thousand Oaks, CA: Sage.

Long, J. Scott. 1997. *Regression Models for Categorical Data*. Thousand Oaks, CA: Sage.

McCullagh, P., and J. A. Nelder. 1989. *Generalized Linear Models*, 2nd ed. London: Chapman and Hall.

McKelvey, Richard D., and William Zavoina. 1975. "A Statistical Model for the Analysis of Ordinal Level Dependent Variables." *Journal of Mathematical Sociology*, 4: 103–120. DOI:10.1080/0022250X.1975.9989847.

Melgar, Natalia, Maximo Rossi, and Tom W. Smith. 2010. "The Perception of Corruption." *International Journal of Public Opinion Research*, 22: 120–131. DOI:10.1093/ijpor/edp058.

Miller, Joanne M., and Jon A. Krosnick. 2000. "News Media Impact on the Ingredients of Presidential Evaluations: Politically Knowledgeable Citizens are Guided by a Trusted Source." *American Journal of Political Science*, 44: 301–315.

Mutz, Diana C., and Paul S. Martin. 2001. "Facilitating Communication Across Lines of Political Difference: The Role of Mass Media." *American Political Science Review*, 95: 97–114.

Norusis, Marija J. 2005. *SPSS 14.0 Advanced Statistical Procedures Companion*. Upper Saddle River, NJ: Prentice Hall.

Powers, Daniel A., and Yu Xie. 2000. *Statistical Methods for Categorical Data Analysis*. San Diego, CA: Academic Press.

Smith, V. Kerry, and William H. Desvousges. 1990. "Risk Communication and the Value of Information: Radon as a Case Study." *Review of Economics and Statistics*, 72: 137–142.

Sommerfeldt, Erich J. 2013. "Online Power Resource Management: Activist Resource Mobilization, Communication Strategy, and Organizational Structure." *Journal of Public Relations Research*, 25: 347–367. DOI:10.1080/1062726X.2013.806871.

Stokes, Maura E., Charles S. Davis, and Gary G. Koch. 2012. *Categorical Data Analysis Using SAS*, 3rd ed. Cary, NC: SAS Institute, Inc.

Tang, Wan, Hua He, and Xin M. Tu. 2012. *Applied Categorical and Count Data Analysis*. Boca Raton: FL: CRC Press.

The American National Election Studies (ANES). American National Election Study, 2008: Pre- and Post-Election Survey. ICPSR25383-v2. Ann Arbor, MI: Inter-university Consortium for Political and Social Research [distributor], 2012-08-30.

Tillema, Taede, Martin Dijst, and Tim Schwanen. 2010. "Face-to-face and Electronic Communications in Maintaining Social Networks: The Influence of Geographical and Relational Distance and of Information Content." *New Media & Society*, 12: 965–983. DOI:10.1177/1461444809353011.

Vigderhous, Gideon. 1977. "Probit Analysis of Radio Ad Awareness." *Journal of Advertising Research*, 17(2): 21–26.

Winship, Christopher, and Robert D. Mare. 1984. "Regression with Ordinal Variables." *American Sociological Review*, 49: 512–525.

Wirth, Michael O., and Harry Bloch. 1989. "Household-Level Demand for Cable Television: A Probit Analysis." *Journal of Media Economics*, 2(2): 21–34. DOI:10.1080/08997768909358183.

10

Poisson and Negative Binomial Regression

Poisson and negative binomial regression techniques model count data as a function of one or more independent variables (Frome 1983, Frome, Kutner, and Beauchamp 1973). Examples of counts, which King (1988, 838) defined as "the number of occurrences of an event in a fixed domain," include the number of news reports published on a certain topic in a two-week period, the number of specific gestures made during a series of debates, and the number of press releases distributed during the management of a crisis. Counts also might include "tweets" sent out by audience members during a major sporting event, the number of individuals "liking" a Facebook page designed to effect some sort of social change, or the number of individuals contacting a public official during a one-month interval. While researchers have used ordinary least squares (OLS) regression to analyze count data, Gardner, Mulvey, and Shaw (1995) explained that OLS procedures often produce negative predicted values when the lower bound of a count is actually zero, and that analyses with count data often violate the OLS assumption of equal variances (see also Tang, He, and Tu 2012, 173–200).

As a generalized linear model, Poisson regression contains a log link function, a Poisson random component, and one or more independent variables as systematic components (Dunteman and Ho 2006). In the Poisson model, explanatory measures may be categorical factors or continuous covariates (Long 1997, 217–230), and the exponentiation of parameter estimates results in *incidence rate ratios*. As the chapter explains, the Poisson distribution contains one parameter, which represents both the mean and the variance; when the variance of a measure exceeds its mean (which is not unusual), data are considered *overdispersed* and in need of a different model.

Negative binomial analyses prove useful when the data in a Poisson model appear overdispersed (DeMaris 2004, Gardner, Mulvey, and Shaw 1995), largely

Categorical Statistics for Communication Research, First Edition. Bryan E. Denham.
© 2017 John Wiley & Sons, Inc. Published 2017 by John Wiley & Sons, Inc.
Companion Website: www.wiley.com/go/denham/categorical_statistics

because the negative binomial model contains an extra parameter that allows the variance of a measure to exceed its mean (Long 1997, 230–238). As Gardner, Mulvey, and Shaw (1995, 399) explained, "The negative binomial can be viewed as a form of Poisson regression that includes a random component reflecting the uncertainty about the true rates at which events occur for individual cases." Because the added parameter can result in a slight loss of power, though, the Poisson model should be used when data are not overdispersed.

The following section reviews communication studies utilizing Poisson and negative binomial regression models. The chapter then addresses the fundamental components of the two models as well as zero-truncated and zero-modified count models, before offering instructions for the SPSS Generalized Linear Models procedure.

Examples of Published Research

Communication scholars have used Poisson and negative binomial regression models to study elements of political and health communication, interpersonal and relational communication, as well as mass communication and society. In recent years the techniques have proven especially useful for studies involving computer-mediated communication, where scholars often examine discrete observations, or counts, across certain periods of time. As an example, in a study involving social media, Choi (2014) used negative binomial analyses to model counts of "retweets." Valenzuela, Arriagada, and Scherman (2012) used a Poisson model in identifying an association between Facebook use and protest activity, while Jiang, Bazarova, and Hancock (2013) analyzed self-disclosure counts in online communication.

In studies of broadcasting, Tsfati, Elfassi, and Waismel-Manor (2010) used Poisson regression in testing the explanatory power of physical attractiveness on the number of times each member of the 16th Israeli Knesset appeared on national television news. Larson and Andrade (2005) used Poisson regression in addressing media coverage of US congresswomen, with explanatory measures including bill sponsorship, media market size, political party, and seniority. Additionally, Yan and Napoli (2006) studied counts of public-affairs programs on broadcast television stations across a two-week period.

In research involving substance abuse, Dasgupta, Mandl, and Brownstein (2009) used Poisson regression in observing an association between news reports and fatal opioid poisonings. Reasoning that community newspapers sometimes function as a macro-level source of social control, Yamamoto and Ran (2013) used Poisson regression and negative binomial models in studying potential links between penetration of community newspapers and number of drug violations. Sorenson, Peterson Manz, and Berk (1998) analyzed determinants of news reports published about homicide victims, and in a study testing cultivation theory, Diefenbach and West (2001) used Poisson regression to analyze estimated crime rates.

Internationally, in a study of Chinese media, Wang (2009) observed significant increases in certain types of articles pertaining to intellectual property. Vliegenthart and Walgrave (2008) examined intermedia agenda-setting processes through a longitudinal analysis of 25 issues covered in Belgian news media, while Himelboim, Chang, and McCreery (2010) studied foreign news coverage in the digital era, observing through Poisson regression a continuation of existing hierarchies. In the context of medicine, Mitchell et al. (2012) used the Poisson model in a study addressing health literacy, with counts of return trips to a hospital serving as a dependent variable. In a study of the heuristic-systematic model, Griffin et al. (2002) used Poisson regression to analyze a dependent variable measuring counts of strongly held beliefs, and in research on group communication, Bazarova and Hancock (2012) studied effects of attributions, using negative binomial analyses to model counts of discussion units.

Methodologically, Tsfati and Peri (2006) analyzed the relationship between mainstream media skepticism and exposure to extranational news media, using the Poisson regression model to manage a severely skewed dependent variable. Analyzing knowledge gaps, Blanks Hindman (2012) also opted for the Poisson model because of problems with skewness and kurtosis. Lastly, researchers have used advanced Poisson techniques to examine communication processes. In research addressing economic news coverage, for instance, Fogarty (2005) used an autoregressive Poisson model, and in studying newspaper coverage of cancer prevention, Slater et al. (2009) developed a multilevel Poisson regression model.

Poisson Regression: Fundamentals

Researchers across the social sciences have analyzed count data with techniques such as ordinary least squares regression, but given the assumptions of parametric statistics and the propensity of linear models to predict negative values, OLS analyses often produce biased (or fundamentally impossible) estimates (DeMaris 2004, 352). Additionally, as Gardner, Mulvey, and Shaw (1995) pointed out, collapsing counts into a limited set of categories may also prove problematic, as collapsing data reduces statistical power and also may result in a loss of information. Moreover, the authors noted, "cut points" in the data may be somewhat arbitrary, eliminating the capacity of analyses to provide reliable answers to research questions. Poisson regression resolves issues associated with OLS models and eliminates the need to collapse or otherwise alter data.

The Poisson distribution itself is an approximation of the binomial distribution and is especially useful for calculating probabilities of rare events in large populations (Nussbaum 2015, 349). In the Poisson distribution, one parameter represents both the mean and the variance, indicating *equidispersion*. The distribution is expressed as:

$$\Pr(Y = y) = \frac{\lambda^{y} e^{-\lambda}}{y!}$$

where y is the number of "hits" (Nussbaum 2015, 349) for a certain time period and λ is the average number of hits for all time periods (see also Dunteman and Ho 2006, 22–23). As an example, one might consider a researcher who monitors news content for 12 weeks and observes 30 articles addressing a certain subject. On average, 2.5 articles appear each week and, if so inclined, the researcher could use the Poisson distribution to calculate the probability that, say, five articles will appear in a given week:

$$\Pr(Y = 5) = \frac{\left(2.5^{5}\right)\left(e^{-2.5}\right)}{5!} = \frac{(97.66)(.082)}{120} = \frac{8.01}{120} = .067$$

Based on the Poisson distribution, the probability of observing five articles in a given week is relatively modest, at .067. Using the same distribution, the probability of two articles appearing in a seven-day period is considerably higher, at .256.

Thus, in the Poisson distribution, λ is the mean, or expected value, and is used in calculating count probabilities. In Poisson regression analysis, one examines potential influences on λ in the form of explanatory measures and potential interactions. Poisson regression models a canonical link, $\ln\lambda$, as a linear function of explanatory variables ($\ln\lambda = \beta_0 + \beta_1 X_1 + \ldots + \beta_p X_p$). The canonical link, used in modeling exponential relationships, ensures non-negative estimates, consistent with a lower bound of zero for count data (Nussbaum 2015, 353). Because of the difficulties associated with interpreting log values, researchers generally exponentiate both sides of an additive log model (shown above) to form a multiplicative model (Dunteman and Ho 2006, 23). Readers may recall this process from previous chapters.

As with other generalized linear models, Poisson regression produces output indicating goodness of fit as well as unstandardized parameter estimates, standard errors, confidence intervals, and significance tests. For each parameter estimate, the Poisson model indicates a Wald chi-square value, and the exponentiation of each estimate produces an incident rate ratio. Hilbe (2011, 15) elaborated on this term through the concept of risk, which, he noted, is central to the modeling of counts. He defined risk as:

> An exposure to the chance or probability of some outcome, typically thought of as a loss or injury … Relative risk is the ratio of the probability of disease for a given

risk factor compared with the probability of disease for those not having the risk factor. It is therefore a ratio of two ratios, and it is often referred to as the *risk ratio*, or, when referencing counts, the *incidence rate ratio*.

Hence, in log-linear modeling and logistic regression analysis, the exponentiation of parameter estimates results in odds ratios, but in Poisson regression, which focuses on the analysis of counts, exponentiation produces incidence rate ratios. The chapter discusses the interpretation of exponentiated parameter estimates in its section on SPSS.

In Poisson (and negative binomial) regression, independence among observations is essential (Frome, Kutner, and Beauchamp 1973). As Hilbe (2011, 2) pointed out, when counts lack independence, assumption violations compromise the Poisson probability distribution function, resulting in overdispersion. As noted earlier in the text, lack of independence among observations also leads to an artificial inflation of the sample, and statistical analyses based on the sample may therefore contain technical flaws.

Scholars who use Poisson regression techniques should also be aware of *offset* measures, which function as explanatory controls. As an example, if a communication scholar opted to study annual increases in electronic newspaper subscriptions over a 20-year period, the researcher might add an offset measure indicating population increases across that time. Otherwise, statistical indicators of growth would disregard the fact that with more people may come more subscriptions. In this regard, Nussbaum (2015, 354) expressed the Poisson regression model as:

$$\lambda_i = s_i \left(e^{\alpha + \beta X_i} \right)$$

where λ_i is the predicted value of Υ_i for ith level of X, and s_i refers to the offset. In many cases, an offset measure is not necessary, and in those instances, s_i is equal to 1.0. Regarding SPSS analyses, Nussbaum (2015, 354) made the important point that a researcher entering an offset measure through the SPSS Generalized Linear Models procedure must create a separate variable for the logarithm of s_i. The chapter now discusses negative binomial regression analysis of count data.

Negative Binomial Regression: Fundamentals

As indicated in the previous section, analyses of count data sometimes result in overdispersion, with the variance of a measure exceeding its mean. When overdispersion occurs, researchers typically replace Poisson regression with the negative binomial model, which is derived from a Poisson-gamma mixture distribution (see Hilbe 2011). The negative binomial distribution, which does not assume equidispersion, gets its name from the fact that, as a discrete probability distribution, it focuses on the number of failures that occur before the *r*th success in a sequence of Bernoulli trials (Nussbaum 2015, 374). As DeMaris (2004, 351) explained, the negative binomial distribution contains two parameters, with *p* representing the probability of a success and *r* representing the desired number of successes. In the following formula, *y* refers to the number of failures encountered before the *r*th success, and *p* represents the probability of a success.

$$f(y \mid r, p) = \frac{(r + y - 1)!}{y!(r-1)!} p^r (1-p)^y$$

for *y* = 0, 1, 2, … Given the above formula, one might consider the administration of a telephone survey the day after a presidential debate. Each student in a research methods course has been trained in survey techniques and is expected to complete four telephone interviews before adjourning for the day. The faculty member supervising the survey knows from past experience that one in three individuals contacted by students will agree to participate. Given that information, what is the probability that a student will complete four interviews in eight calls to prospective respondents?

$$f(8 \mid r = 4, p = .33) = \frac{(4 + 8 - 1)!}{8!(4 - 1)!} .33^4 (1 - .33)^8$$

$$\frac{11!}{8!(3)!} .33^4 (.67)^8 = \frac{39916800}{40320(6)} .0119(.0406) = 165(.000483) = .079$$

A negative binomial analysis indicates that a student has less than an 8% chance of completing four interviews in eight attempts given a 33% success rate.

Regarding overdispersion in the negative binomial model, DeMaris (2004, 351) advised inspecting the mean, or expected value of *y*, and the variance, given the following formulas:

$$E(y) = r\frac{1-p}{p} = 4\frac{1-.33}{.33} = 8.12$$

$$V(y) = r\frac{1-p}{p^2} = 4\frac{1-.33}{.33^2} = 24.6$$

In this case, as the negative binomial model would assume, the variance of 24.6 clearly exceeds the mean of 8.12 for the data involving student survey research.

In terms of expression, negative binomial regression takes the same general form as its Poisson counterpart (ln $\lambda = \beta_0 + \beta_1 X_1 + \ldots + \beta_p X_p$), and to facilitate interpretation, researchers generally exponentiate log values to form a multiplicative model. Negative binomial regression is a type of generalized linear model, and like Poisson regression, it is characterized by a log link function as well as a systematic component consisting of categorical and/or continuous explanatory variables. The random component is the negative binomial, which, as indicated, is derived from a Poisson–gamma mixture distribution (see Hilbe 2011). As with Poisson regression, offset measures can be entered to prevent spurious results. Before addressing SPSS techniques for Poisson and negative binomial analyses, the chapter summarizes additional models for count data.

Additional Techniques

In analyses of count data, observations sometimes become part of a sample only after the first count occurs. As an example, Long (1997, 239) noted that if one derived a sample of scientists based on individuals who had published their research in a certain journal, those who had not published their work in the journal would be excluded automatically. Tang, He, and Tu (2012, 196) noted that studies focusing on length of hospitalization for a certain disease typically include only those patients who have been hospitalized; individuals not admitted for treatment generally do not become part of the sample. Statisticians characterize such samples (i.e., those that do not contain the value zero) as *zero truncated*. In such cases, Poisson and negative binomial models, as discussed in the present chapter, are not appropriate; however, extensions of the models do exist, and Long (1997) recommended the work of Grogger and Carson (1991) as well as Gurmu and Trivedi (1992) as starting points.

Situations also arise when zeros dominate sample distributions. Discussing *zero modified count models*, Tang, He, and Tu (2012, 190) referred to a sexual-health distribution indicating frequency of protected vaginal sex. The distribution contained an excessive number of zeros, likely reflecting the fact that many individuals in the sample abstained from sex or practiced unprotected vaginal sex. In another context, Long (1997, 242) addressed the publication of scientific papers. While the Poisson and negative binomial models would assume positive probabilities for all scientists in the publication realm, many individuals work in environments where publication is not a possibility. As Long (1997, 242) explained, "The probability differs across individuals according to their characteristics, but *all* scientists are at risk of not publishing and *all* scientists might publish." Given differing expectations, Long explained that this assumption is not appropriate, and a researcher would want to consider a *zero inflated model* (Greene 1994, Lambert 1992), which Long discussed in some detail.[1]

The following section reviews output statistics for regression analyses of count data, as calculated in the SPSS Generalized Linear Models procedure.

SPSS Analyses

This section of the chapter uses data from the 2008 American National Election Study (The American National Election Studies 2008) to demonstrate both Poisson and negative binomial regression techniques in SPSS. In the examples, a dependent count variable indicates how many days in the previous week respondents ($N = 1,152$) watched national television news, with options ranging from 0 to 7. With this measure, no values below zero can emerge, and the discrete units represent "exposures" in a certain time period. Explanatory measures include sex, race, and attitudes toward the United States (i.e., Optimistic, Pessimistic, Neither Optimistic nor Pessimistic).

In SPSS, the Generalized Linear Models procedure fits both Poisson and negative binomial regression models; to choose between the two, a researcher must determine whether a dependent measure is overdispersed. To check dispersion, one can select Analyze > Descriptive Statistics > Explore and receive a descriptive report similar to the one shown in Table 10.1. The statistics shown in this table indicate that, on average, individuals watched the national television news 3.62 days in the previous week, with a variance of 7.543. These values indicate the data are overdispersed and that a negative binomial model is preferred over a Poisson analysis. For didactic purposes, the chapter covers the SPSS Poisson model first.

Recalling Figure 9.1 from the previous chapter, a Poisson regression analysis can be conducted through the SPSS Generalized Linear Models procedure (Analyze > Generalized Linear Models > Generalized Linear Models). To begin,

Table 10.1 SPSS output for Explore analysis of television news exposure

Descriptives			*Statistic*	*SE*
Days Past Week Watched National News on TV	Mean		3.62	.081
	95% Confidence Interval for Mean	Lower Bound	3.46	
		Upper Bound	3.78	
	5% Trimmed Mean		3.63	
	Median		3.00	
	Variance		7.543	
	Standard Deviation		2.746	
	Minimum		0	
	Maximum		7	
	Range		7	
	Interquartile Range		6	
	Skewness		.008	.072
	Kurtosis		−1.558	.144

one should select "Poisson log-linear" under the Type of Model tab and then enter a dependent variable under Response as well as explanatory measures under Predictors. In the Model window, one can specify a main effects or interaction model and then proceed to the remaining tabs: Estimation, Statistics, EM Means, Save, and Export. As indicated in Chapter 9, the default options under these tabs function reasonably well, but researchers may wish to select additional statistics for the output. As an example, under the Save tab, one can examine dispersion by selecting standardized residuals and taking note of whether the values fall between −3.0 and +3.0. If residuals are consistently scattered beyond these lower and upper bounds, the model selected may need to be replaced with a better-fitting one.

Table 10.2 displays the results of a Poisson regression analysis. Model Information confirms (a) the dependent variable of exposure to national television news, (b) the Poisson probability distribution, and (c) the log link function. Following that information, a series of goodness-of-fit statistics appear, with both the Deviance and Pearson Chi-Square values indicating overdispersed data. Ideally, ratios of item values and their respective degrees of freedom will appear close to 1.0. In the current analysis, the Deviance Value/df ratio of 2.572 and the Pearson Chi-Square ratio of 1.993 suggest overdispersion. One can also observe a value for the full log-likelihood function in addition to other goodness-of-fit indicators, notably the Bayesian Information Criterion (BIC). Generally, a better-fitting model will contain a comparably low BIC value.

Next, under Omnibus Test, the value for Likelihood Ratio Chi-Square (62.525) is statistically significant, indicating that a model containing the three explanatory measures fits better than a model containing only the intercept. In the section titled Test of Model Effects, the explanatory power of each predictor is tested independently of the others, and looking at the results, it appears both race and attitude toward the United States showed significance. On the other hand, sex did not appear significant. The statistics shown under Parameter Estimates reveal that with sex and attitude toward the United States included in the regression equation, race did not function as a significant predictor. Only the attitudinal item showed significant explanatory power.

As shown in Table 10.2, parameter estimates for those who indicated optimism (.266) as well as pessimism (.259) were positive and significant at $p < .001$, with both estimates exponentiating to approximately 1.30. In terms of incidence rate ratios, the exponentiated estimates reveal that individuals who were optimistic as well as those who were pessimistic watched national television news at higher rates than individuals who were neither optimistic nor pessimistic. An exponentiated estimate of 1.30 indicates that rates increased about 30%, on average, among optimistic and pessimistic individuals, given the presence of sex and race in the regression equation.

Table 10.3 displays the results of a negative binomial regression analysis. This model offered a better fit to the survey data, as indicated by the goodness-of-fit

Table 10.2 SPSS output for Poisson regression model in Generalized Linear Models procedure

Model Information	
Dependent Variable	Days past week watched national news on TV
Probability Distribution	Poisson
Link Function	Log

Goodness of Fit[b]	Value	df	Value/df
Deviance	2626.188	1021	2.572
Scaled Deviance	2626.188	1021	
Pearson Chi-Square	2035.040	1021	1.993
Scaled Pearson Chi-Square	2035.040	1021	
Log Likelihood[a]	−2653.478		
Akaike's Information Criterion (AIC)	5318.957		
Finite Sample Corrected AIC (AICC)	5319.039		
Bayesian Information Criterion (BIC)	5348.563		
Consistent AIC (CAIC)	5354.563		

[a] The full log likelihood function is displayed and used in computing information criteria.
[b] Information criteria are in small-is-better form.

Omnibus Test[a]		
Likelihood Ratio Chi-Square	df	Sig.
62.525	5	.000

[a] Compares the fitted model against the intercept-only model.

Test of Model Effects

	Type III		
Source	Wald Chi-Square	df	Sig.
Intercept	3307.737	1	.000
Sex	.701	1	.403
Race	11.230	2	.004
Attitude about US	48.062	2	.000

Parameter Estimates

			95% Wald Confidence Interval		Hypothesis Test		
Parameter	B	SE	Lower	Upper	Wald Chi-Square	df	Sig.
Intercept	1.111	.0558	1.001	1.220	396.528	1	.000
Male	.027	.0328	−.037	.092	.701	1	.403
White	−.050	.0518	−.151	.052	.923	1	.337
Black	.076	.0561	−.034	.186	1.824	1	.177
Optimistic	.266	.0393	.188	.343	45.636	1	.000
Pessimistic	.259	.0534	.155	.364	23.569	1	.000
(Scale)	1[a]						

[a] Fixed at the displayed value.

Table 10.3 SPSS output for negative binomial model in Generalized Linear Models procedure

Model Information

Dependent Variable	Days past week watched national news on TV
Probability Distribution	Negative binomial
Link Function	Log

Goodness of Fit[b]

	Value	df	Value/df
Deviance	860.345	1021	.843
Scaled Deviance	860.345	1021	
Pearson Chi-Square	444.744	1021	.436
Scaled Pearson Chi-Square	444.744	1021	
Log Likelihood[a]	−2491.699		
Akaike's Information Criterion (AIC)	4995.398		
Finite Sample Corrected AIC (AICC)	4995.481		
Bayesian Information Criterion (BIC)	5025.005		
Consistent AIC (CAIC)	5031.005		

[a]The full log likelihood function is displayed and used in computing information criteria.
[b]Information criteria are in small-is-better form.

Omnibus Test

Likelihood Ratio Chi-Square	df	Sig.
13.919	5	.016

Test of Model Effects

Type III

Source	Wald Chi-Square	df	Sig.
Intercept	694.121	1	.000
Sex	.139	1	.709
Race	2.602	2	.272
Attitude about US	11.603	2	.003

Parameter Estimates

Parameter	B	SE	95% Wald Confidence Interval Lower	95% Wald Confidence Interval Upper	Hypothesis Test Wald Chi-Square	df	Sig.
Intercept	1.108	.1179	.876	1.339	88.179	1	.000
Male	.027	.0715	−.113	.167	.139	1	.709
White	−.052	.1124	−.272	.169	.212	1	.645
Black	.082	.1230	−.159	.323	.442	1	.506
Optimistic	.269	.0816	.109	.429	10.877	1	.001
Pessimistic	.266	.1141	.042	.490	5.427	1	.020
(Scale)	1[a]						
(Neg. binomial)	1						

[a]Fixed at the displayed value.

statistics. The Deviance Value/df ratio of .843 approached the ideal estimate of 1.0, although the Pearson Chi-Square ratio of .436 was not as close. In the Omnibus Test, the Likelihood Ratio Chi-Square value of 13.919 showed significance at $p < .05$, indicating that a model containing the three explanatory measures fit better than a model containing only the intercept. Examining the Test of Model Effects, the negative binomial regression differed from the Poisson analysis in that the former did not report significant effects for race. Only the attitudinal measure contributed significantly and that pattern held in the full model, where sex and race did not show significance. As in the Poisson analysis, individuals who expressed optimism as well as those who expressed pessimism appeared to watch national television news at higher rates than survey respondents who indicated neither optimism nor pessimism. The parameter estimates .269 and .266 exponentiated to approximately 1.30, again suggesting higher rates of viewership.

Examining Table 10.3, one also observes a value for the dispersion coefficient in the negative binomial model. In the Poisson model, this coefficient is set to zero and is not reported. When this item is not constrained (as it is in the Poisson model), a value greater than zero indicates overdispersion. In rare cases, a value less than zero will emerge and indicate underdispersion, where the variance is less than the mean. In this case the value of 1.0 confirms that a negative binomial model is the preferred technique for analyzing predictors of the number of days respondents reported watching national television news.

Chapter Summary

This chapter has addressed Poisson and negative binomial regression, two techniques used in analyzing count data. Poisson regression assumes equidispersion while the negative binomial model assumes overdispersion and includes an extra parameter. Explanatory measures for both types of regression may be categorical factors or continuous covariates, and exponentiated parameter estimates produce incidence rate ratios for the categories of predictor variables. In some cases, Poisson and negative binomial models should include offset measures, as counts may vary as a result of factors not accounted for among explanatory variables. Additionally, certain situations may call for extended versions of the Poisson and negative binomial regression models.

Chapter Exercises

1. Violence against women is a significant social problem in the United States, and researchers have studied the issue from the standpoint of communication (see Weathers and Hopson 2015, Yokotani 2015). The table below contains state data indicating the number of women killed by men in 2012 (Violence Policy Center 2014). The most homicides per 100,000 women

occurred in the 10 states listed first, while the fewest took place in the 10 states listed in the second group. Also listed is the female population for each state as well as the region of the country assigned to each state by the US Census Bureau. Using the Explore function in SPSS, determine whether a Poisson or negative binomial model would be the appropriate choice for analyzing the explanatory power of region on counts of female homicides. After making that determination, fit a model to the data, first without the female population figures and then with the figures entered as a covariate.[2] Note the differences in the two models as you report the findings. For the model that includes the covariate, provide goodness-of-fit statistics, results of the omnibus test, as well as tests of model effects. Also include parameter estimates, standard errors, 95% confidence intervals, and the results of hypothesis tests, as shown in SPSS output. Did counts vary by region?

2. Using additional data in the table, repeat the steps taken in the previous question and test the explanatory power of region on the number of aggravated assaults committed (against members of both sexes) in each state in 2012.[3] Test the effects of region on its own and then include state populations as a covariate. Report all of the results listed in the previous question and interpret the findings on a substantive level.

State	Region	State Population	Female Population	Female Homicides	Aggravated Assaults
Alaska	9	731,449	351,096	9	3,169
South Carolina	5	4,723,723	2,427,993	50	19,905
Oklahoma	7	3,814,820	1,926,484	39	12,867
Louisiana	7	4,601,893	2,346,965	45	15,740
Mississippi	6	2,984,926	1,534,252	29	4,460
Nevada	8	2,758,931	1,365,670	25	10,790
Missouri	4	6,021,988	3,071,213	53	19,473
Arizona	8	6,553,255	3,296,287	56	18,087
Georgia	5	9,919,945	5,079,011	84	22,423
Tennessee	6	6,456,243	3,312,052	53	30,961
Minnesota	4	5,379,139	2,711,086	19	7,207
Utah	8	2,855,287	1,421,943	10	3,783
Connecticut	1	3,590,347	1,841,848	12	5,408
Hawaii	9	1,392,313	694,764	9	1,976
Iowa	4	3,074,186	1,552,463	9	6,234
Massachusetts	1	6,646,144	3,429,410	17	18,638
Vermont	1	626,011	318,013	1	652
Nebraska	4	1,855,525	935,184	3	2,920
New Hampshire	1	1,320,718	669,604	2	1,545
Illinois	3	12,875,255	6,552,504	16	29,618

Note: Excel file containing data available on companion website.

Notes

1 Zero inflated models resemble the *tobit* model (Tobin 1958) insofar as both assume a cluster of nonnegative values, usually zero (McDonald and Moffitt 1980). However, the tobit model assumes an underlying continuous distribution, and although studies of count data have occasionally used this model, Poisson and negative binomial analyses, and their respective extensions, are more appropriate for discrete counts.
2 In SPSS, converting the populations to standardized Z-scores and using the scores as a covariate will facilitate regression output. Using raw figures may result in undetermined estimates. To establish Z-scores in SPSS, click on Descriptives and enter the variables of interest. Then check the box in the lower part of the window to request the calculation of standardized scores. A new column of Z-scores will appear in the data window.
3 Data obtained from the FBI Uniform Crime Report, available at http://www.fbi.gov/about-us/cjis/ucr/crime-in-the-u.s/2012/crime-in-the-u.s.-2012/tables/5tabledatadecpdf. State populations based on 2012 US Census, available at http://www.governing.com/gov-data/state-census-population-migration-births-deaths-estimates.html.

References

Bazarova, Natalya N., and Jeffrey T. Hancock. 2012. "Attributions After a Group Failure: Do They Matter? Effects of Attributions on Group Communication and Performance." *Communication Research*, 39: 499–512. DOI:10.1177/0093650210397538.

Blanks Hindman, Douglas. 2012. "Knowledge Gaps, Belief Gaps, and Public Opinion about Health Care Reform." *Journalism & Mass Communication Quarterly*, 89: 585–605. DOI: 10.1177/1077699012456021.

Choi, Sujin. 2014. "Flow, Diversity, Form, and Influence of Political Talk in Social-Media-Based Public Forums." *Human Communication Research*, 40: 209–237. DOI:10.1111/hcre.12023.

Dasgupta, Nabarun, Kenneth D. Mandl, and John S. Brownstein. 2009. "Breaking the News or Fueling the Epidemic? Temporal Association Between News Media Report Volume and Opioid-Related Mortality." *PLoS ONE*, 4(11): e7758. DOI:10.1371/journal.pone.0007758.

DeMaris, Alfred. 2004. *Regression with Social Data: Modeling Continuous and Limited Response Variables.* Hoboken, NJ: John Wiley & Sons.

Diefenbach, Donald L., and Mark D. West. 2001. "Violent Crime and Poisson Regression: A Method and a Measure for Cultivation Analysis." *Journal of Broadcasting and Electronic Media*, 45: 432–445. DOI:10.1207/s15506878jobem4503_4.

Dunteman, George H., and Moon-Ho R. Ho. 2006. *An Introduction to Generalized Linear Models.* Thousand Oaks, CA: Sage.

Fogarty, Brian J. 2005. "Determining Economic News Coverage." *International Journal of Public Opinion Research*, 17: 149–172. DOI:10.1093/ijpor/edh051.

Frome, E. L. 1983. "The Analysis of Rates Using Poisson Regression Models." *Biometrics*, 39: 665–674.

Frome, Edward L., Michael H. Kutner, and John J. Beauchamp. 1973. "Regression Analysis of Poisson-Distributed Data." *Journal of the American Statistical Association*, 68: 935–940. DOI:10.1080/01621459.1973.10481449.

Gardner, William, Edward P. Mulvey, and Esther C. Shaw. 1995. "Regression Analyses of Counts and Rates: Poisson, Overdispersed Poisson, and Negative Binomial Models." *Psychological Bulletin*, 118: 392–404. DOI:10.1037/0033-2909.118.3.392.

Greene, William H. 1994. "Accounting for Excess Zeros and Sample Selection in Poisson and Negative Binomial Regression Models." Working Paper No. 94-10. New York: Stern School of Business, New York University, Department of Economics.

Griffin, Robert J., Kurt Neuwirth, James Giese, and Sharon Dunwoody. 2002. "Linking the Heuristic-Systematic Model and Depth of Processing." *Communication Research*, 29: 705–732. DOI:10.1177/009365002237833.

Grogger, J. T., and R. T. Carson. 1991. "Models for Truncated Counts." *Journal of Applied Econometrics*, 6: 225–238. DOI:10.1002/jae.3950060302.

Gurmu, Shiferaw, and Pravin K. Trivedi. 1992. "Overdispersion Tests for Truncated Poisson Regression Models." *Journal of Econometrics*, 54: 347–370. DOI:10.1016/0304-4076(92)90113-6.

Hilbe, Joseph M. 2011. *Negative Binomial Regression*, 2nd ed. Cambridge: Cambridge University Press.

Himelboim, Itai, Tsan-Kuo Chang, and Stephen McCreery. 2010. "International Network of Foreign News Coverage: Old Global Hierarchies in a New Online World." *Journalism & Mass Communication Quarterly*, 87: 297–314. DOI:10.1177/1077699901008700205.

Jiang, L. Crystal, Natalya N. Bazarova, and Jeffrey T. Hancock. 2013. "From Perception to Behavior: Disclosure Reciprocity and the Intensification of Intimacy in Computer-Mediated Communication." *Communication Research*, 40: 125–143. DOI:10.1177/0093650211405313.

King, Gary. 1988. "Statistical Models for Political Science Event Counts: Bias in Conventional Procedures and Evidence for the Exponential Poisson Regression Model." *American Journal of Political Science*, 32: 838–863.

Lambert, Diane. 1992. "Zero-inflated Poisson Regression with an Application to Defects in Manufacturing." *Technometrics*, 34: 1–14. DOI:10.1080/00401706.1992.10485228.

Larson, Stephanie Greco, and Lydia M. Andrade. 2005. "Determinants of National Television News Coverage of Women in the House of Representatives, 1987–1998." *Congress & the Presidency*, 32: 49–61. DOI:10.1080/07343460509507696.

Long, J. Scott. 1997. *Regression Models for Categorical and Limited Dependent Variables*. Thousand Oaks, CA: Sage.

McDonald, John F., and Robert A. Moffitt. 1980. "The Uses of Tobit Analysis." *The Review of Economics and Statistics*, 62: 318–321.

Mitchell, Suzanne E., Ekaterina Sadikova, Brian W. Jack, and Michael K. Paasche-Orlow. 2012. "Health Literacy and 30-day Postdischarge Hospital Utilization." *Journal of Health Communication*, 17: 325–338. DOI:10.1080/10810730.2012.715233.

Nussbaum, E. Michael. 2015. *Categorical and Nonparametric Data Analysis*. New York: Routledge.

Slater, Michael D., Andrew F. Hayes, Jason B. Reineke, Marilee Long, and Edwin P. Bettinghaus. 2009. "Newspaper Coverage of Cancer Prevention: Multilevel Evidence for Knowledge-Gap Effects." *Journal of Communication*, 59: 514–533. DOI:10.1111/j.1460–2466.2009.01433.x.

Sorenson, Susan B., Julie G. Peterson Manz, and Richard A. Berk. 1998. "News Media Coverage and the Epidemiology of Homicide." *American Journal of Public Health*, 88: 1510–1514. DOI: 10.2105/AJPH.88.10.1510.

Tang, Wan, Hua He, and Xin M. Tu. 2012. *Applied Categorical and Count Data Analysis*. Boca Raton, FL: CRC Press.

The American National Election Studies. 2008. American National Election Study: ANES Pre- and Post-Election Survey. ICPSR25383-v2. Ann Arbor, MI: Inter-university Consortium for Political and Social Research [distributor], 2012–08–30. DOI:10.3886/ICPSR25383.v2.

Tobin, James. 1958. "Estimation of Relationships for Limited Dependent Variables." *Econometrica*, 26: 24–36.

Tsfati, Yariv, Dana Markowitz Elfassi, and Israel Waismel-Manor. 2010. "Exploring the Association Between Israeli Legislators' Physical Attractiveness and Their Television News Coverage." *International Journal of Press/Politics*, 15: 175–192. DOI:10.1177/1940161209361212.

Tsfati, Yariv, and Yoram Peri. 2006. "Mainstream Media Skepticism and Exposure to Sectorial and Extranational News Media: The Case of Israel." *Mass Communication & Society*, 9: 165–187. DOI:10.1207/s15327825mcs0902_3.

Valenzuela, Sebastian, Arturo Arriagada, and Andres Scherman. 2012. "The Social Media Basis of Youth Protest Behavior: The Case of Chile." *Journal of Communication*, 62: 299–314. DOI:10.1111/j.1460-2466.2012.01635.x.

Violence Policy Center. 2014. *When Men Murder Women: An Analysis of the 2012 Homicide Data*. Foundation report available at http://www.vpc.org/studies/wmmw2014.pdf.

Vliegenthart, Rens, and Stefaan Walgrave. 2008. "The Contingency of Intermedia Agenda Setting: A Longitudinal Study in Belgium." *Journalism & Mass Communication Quarterly*, 85: 860–877. DOI:10.1177/107769900808500409.

Wang, Mian R. 2009. "An Analysis of Intellectual Property Discourse in Chinese Media." *China Media Research*, 5: 64–74.

Weathers, Melinda, and Mark C. Hopson. 2015. "'I Define What Hurts Me': A Co-Cultural Theoretical Analysis of Communication Factors Related to Digital Dating Abuse." *Howard Journal of Communications*, 26: 95–113.

Yamamoto, Masahiro, and Weina Ran. 2013. "Drug Abuse Violations in Communities: Community Newspapers as a Macro-level Source of Social Control." *Journalism & Mass Communication Quarterly*, 90: 629–651. DOI:10.1177/1077699013503164.

Yan, Michael Zhaoxu, and Philip M. Napoli. 2006. "Market Competition, Station Ownership, and Local Public Affairs Programming on Broadcast Television." *Journal of Communication*, 56: 795–812. DOI:10.1111/j.1460–2466.2006.00320.x.

Yokotani, Kenji. 2015. "Links between Impolite Spousal Forms of Address and Intimate Partner Violence Against Women." *Journal of Language & Social Psychology*, 34: 213–221.

11

Interrater Agreement Measures for Nominal and Ordinal Data

In addition to conducting experiments and administering surveys, communication scholars often examine the content of media texts and conversation transcripts by quantifying occurrences of certain terms or characteristics. Bernard Berelson (1952, 18) described this type of investigation, content analysis, as "a research technique for the objective, systematic, and quantitative description of the manifest content of communication." To conduct an objective and systematic study of communication, a researcher must define content measures as precisely as materials allow. For truly objective variables, such as the name of a newspaper or the date of an article, coding classifications are relatively straightforward. But when measures become subjective, calling for interpretation, assigning numbers to content becomes more difficult. In such cases, the precision of operational definitions will determine the extent to which coding processes prove reliable.

This chapter focuses on measures of interrater agreement, which researchers use to assess reliability in content analyses. Lacking indications of agreement, scholars cannot determine whether coding procedures will result in consistent content observations. As Hayes and Krippendorff (2007, 78) asked, "Are the data being made and subsequently used in analyses and decision making the result of irreproducible human idiosyncrasies or do they reflect properties of the phenomena (units of analysis) of interest on which others could agree as well?" As an example, scholars often study valence, or implicit degree of attractiveness, in visual images, coding content as positive, negative, or neutral (see Lang et al. 2011, Rodriguez and Asoro 2012). In political communication, a researcher might be interested in whether photographs of a public official vary in valence across news outlets, or whether valence changes before and after a pivotal event. Prior to analyzing content, a researcher would need to operationalize Positive, Negative, and Neutral as content categories and then test interrater agreement to ensure consistent coding (see, for discussion, Hayes 2005, 118–129, Krippendorff 1980, 129–154, Popping 2010, Potter and Levine-Donnerstein 1999, Riffe, Lacy, and Fico 2005, 123–159).[1]

Although reliability is central to content analysis, research has found that communication scholars, in general, do not provide sufficient information about agreement in published studies (see Lombard, Snyder-Duch, and Bracken 2002). Some studies report no information about agreement, while others report basic indicators of percentage agreement and percentage-agreement equivalents such as Holsti's (1969) formula. A fundamental problem with these indicators is that they do not account for chance agreement among coders (see Mitchell 1979, Krippendorff 2004, 2008, 2011), and depending on the distribution of coded material, accounting for chance can prove integral to a content analysis.

Analysis of Nominal Data with Two Raters

This section of the chapter reviews three measures of interrater agreement for nominal data: Cohen's kappa, Scott's pi, and Krippendorff's alpha. Calculations for these measures, each of which accounts for chance agreement among two raters, will be demonstrated based on data contained in Table 11.1. In a hypothetical study, two researchers have independently coded 83 news photographs for valence, classifying each image as positive, negative, or neutral. Numbers along the diagonal (in bold) indicate agreement on $3 + 51 + 3 = 57$ (68.6%) of 83 items. But a large number of ratings appear in a single cell, with coders agreeing on negative imagery in 51 (61.4%) of 83 instances. Such an imbalance can increase chance agreement, and thus, in this case, the researchers have two reasons for testing interrater reliability: the subjectivity of the valence measure as well as data distribution.

Table 11.1 Cell frequencies for interrater reliability calculations

		Rater 1			
		Positive	*Negative*	*Neutral*	*Marginal*
Rater 2	Positive	3	6	4	13
	Negative	4	**51**	3	58
	Neutral	3	6	**3**	12
	Marginal	10	63	10	83

The fundamental task in calculating all three measures of interrater reliability is to compare observed (P_o) and expected (P_e) proportions of agreement, as shown in the following equation from Scott (1955), whose pi measure appeared prior to kappa and alpha in the scholarly literature:

$$\frac{P_o - P_e}{1 - P_e}$$

Applying Scott's general formula to the data in Table 11.1, proportion of observed agreement has already been established ($P_o = .686$). The next task is to calculate expected proportion of agreement, where n refers to the number of cases each researcher rates, r_i to the marginal frequency total for the ith row, c_i to the marginal frequency total for the ith column, and p_i to the expected proportion in the ith cell along the diagonal, such that $P_e = \Sigma p_i$. Of the three interrater reliability measures discussed in this section, readers may find computations for kappa the most intuitive, primarily because calculations for the kappa statistic resemble those used in chi-square analysis, covered in Chapter 2.

Cohen's Kappa

Statisticians generally consider kappa (Cohen 1960) the most popular measure of agreement for categorical data (Agresti 2007, 264, Tang, He, and Tu 2012, 57). Its popularity is due, in part, to its availability in software packages such as SPSS. Kappa accounts for chance agreement and varies between -1 and $+1$, where $+1$ equals perfect agreement, 0 equals chance agreement, and a negative number indicates less-than-chance agreement. Kappa assumes independence among coders as well as units of analysis, and it requires variable categories to be mutually exclusive and exhaustive.

In estimating an expected proportion for each diagonal cell in a cross-tabulation, kappa uses each coder's frequencies to estimate each coder's expected proportion, weighting one with the other to arrive at an expected value. For kappa, each coder's expected proportion is taken as the number of times the coder judged a unit to be in a particular category divided by the number of judgments. Thus:

$$p_i = \left(\frac{r_i}{n}\right)\left(\frac{c_i}{n}\right) = \frac{r_i c_i}{n^2}$$

Recalling that $P_e = \Sigma p_i$, the following calculations produce expected proportions for Table 11.1:

$$\left(\frac{13}{83}\right)\left(\frac{10}{83}\right) = \frac{(13)(10)}{83^2} = \frac{130}{6,889} = .019$$

$$+$$

$$\left(\frac{58}{83}\right)\left(\frac{63}{83}\right) = \frac{(58)(63)}{83^2} = \frac{3,654}{6,889} = .530$$

$$+$$

$$\left(\frac{12}{83}\right)\left(\frac{10}{83}\right) = \frac{(12)(10)}{83^2} = \frac{120}{6,889} = .017$$

In this case, $P_e = .019 + .530 + .017 = .566$. One can then calculate Cohen's kappa:

$$\frac{P_o - P_e}{1 - P_e} = \frac{.686 - .566}{1 - .566} = \frac{.12}{.434} = .276$$

In their frequently cited study addressing agreement with categorical data, Landis and Koch (1977a) established the following "benchmarks" for the kappa coefficient:

< 0.00	Poor
0.00–0.20	Slight
0.21–0.40	Fair
0.41–0.60	Moderate
0.61–0.80	Substantial
0.81–1.00	Almost Perfect

Given those benchmarks, .276 would be considered "fair." In this hypothetical situation, a modest kappa value was expected given that agreement on negative images accounted for more than 60% of all cases. Such an imbalance affects marginal totals, used in establishing expected cell proportions. In fact, one criticism of kappa (and pi) is that values can be influenced substantially by marginal distributions (Agresti 2007, 264, Banerjee et al. 1999, 6, Brennan and Prediger 1981); in some cases, marginal counts result in overly conservative estimates (Neuendorf 2002, 151).[2] Additionally, as a reminder, Scott (1955) and Cohen (1960) developed pi and kappa, respectively, for situations in which (only) two raters intended to evaluate (only) nominal data (see, for discussion, Craig 1981, Fleiss, Levin, and Paik 2003, 598–626, Lawal 2003, 483–494, Stokes, Davis, and Koch 2012, 131–134).[3]

Scott's Pi

Recognizing the inherent limitations of simple percentage agreement, Scott (1955) developed pi to account for chance agreement among raters. Like kappa, pi ranges between -1 and $+1$, with 1 showing excellent agreement, 0 representing agreement by chance, and a negative value indicating less-than-chance agreement. In calculations for Scott's pi, each coder's expected proportion is estimated as the average of the two expected proportions computed for Cohen's kappa. Expected proportions for Scott's pi are calculated as follows:

$$p_i = \left(\frac{r_i + c_i}{2n}\right)\left(\frac{r_i + c_i}{2n}\right) = \left(\frac{r_i + c_i}{2n}\right)^2$$

In terms of statistical reasoning, readers who have completed a course in finite mathematics are undoubtedly familiar with widgets and urns. In the present case, suppose that, for each unit judged, each coder labels a widget with the name of a category, and the coders then place all labeled widgets in an urn. The probability of drawing a widget labeled category i equals the number of widgets labeled i (i.e., $r_i + c_i$) divided by the number of widgets ($2n$). Thus, if one draws a widget labeled with a category, returns it to the urn, and then draws a second widget, the probability of both widgets indicating category i is precisely the probability shown in the above equation. Scott's pi probabilities for each category are therefore equivalent to sampling with replacement. Applying the formula to data in Table 11.1, the following calculations produce an expected proportion:

$$\left(\frac{13+10}{(2)(83)}\right)\left(\frac{13+10}{(2)(83)}\right) = \left(\frac{13+10}{(2)(83)}\right)^2 = \left(\frac{23}{166}\right)^2 = .019$$

$$+$$

$$\left(\frac{58+63}{(2)(83)}\right)\left(\frac{58+63}{(2)(83)}\right) = \left(\frac{58+63}{(2)(83)}\right)^2 = \left(\frac{121}{166}\right)^2 = .531$$

$$+$$

$$\left(\frac{12+10}{(2)(83)}\right)\left(\frac{12+10}{(2)(83)}\right) = \left(\frac{12+10}{(2)(83)}\right)^2 = \left(\frac{22}{166}\right)^2 = .018$$

In this case, $P_e = .019 + .530 + .018 = .568$. One can then calculate Scott's pi:

$$\frac{P_o - P_e}{1 - P_e} = \frac{.686 - .568}{1 - .568} = \frac{.118}{.432} = .273$$

For the data in Table 11.1, the value of pi is .273, an estimate close to the value for kappa (.276).

Krippendorff's Alpha

In communication, Lombard, Snyder-Duch, and Bracken (2004) called for a standard measure of reliability in content analysis. Their suggestion appeared shortly after their 2002 review of content studies showed inconsistencies in reliability reporting. Hayes and Krippendorff (2007) responded to Lombard, Snyder-Duch, and Bracken (2004) by proposing that alpha, an agreement measure developed by Krippendorff (1980), serve as the standard.[4] Alpha differs from kappa and pi in that disagreements – not agreements – are of primary interest to the researcher. To accommodate direct comparisons with kappa and pi, however, the chapter transforms alpha to Scott's general formula for calculating pi.[5] As indicated in the formula below, alpha probabilities are equivalent to sampling *without* replacement. In the formula, the first proportion is the same as one of the proportions used in calculating pi, and for the second proportion, one less widget appears in the urn; because it applies to category i, the number of widgets so labeled is also reduced by one. The equation indicates that as n grows large, the value for alpha will approach pi, but when n is small, alpha can be expected to be smaller than pi.

$$p_i = \left(\frac{r_i + c_i}{2n} \right) \left(\frac{r_i + c_i - 1}{2n - 1} \right)$$

For the data in Table 11.1, expected proportions for alpha would be calculated in the following manner:

$$\left(\frac{13 + 10}{166} \right) \left(\frac{13 + 10 - 1}{166 - 1} \right) = \left(\frac{23}{166} \right) \left(\frac{22}{165} \right) = (.139)(.133) = .018$$

$$+$$

$$\left(\frac{58 + 63}{166} \right) \left(\frac{58 + 63 - 1}{166 - 1} \right) = \left(\frac{121}{166} \right) \left(\frac{120}{165} \right) = (.729)(.727) = .530$$

$$+$$

$$\left(\frac{12 + 10}{166} \right) \left(\frac{12 + 10 - 1}{166 - 1} \right) = \left(\frac{22}{166} \right) \left(\frac{21}{165} \right) = (.133)(.127) = .017$$

In this case, $P_e = .018 + .530 + .017 = .565$. One can then calculate Krippendorff's alpha:

$$\frac{.686 - .565}{1 - .565} = \frac{.121}{.435} = .278$$

In this case, the Krippendorff alpha coefficient is a modest .278. Discussing alpha, Krippendorff (1980, 147) suggested that social scientists recognize alpha reliability estimates of .80 or greater as acceptable, with estimates between .67 and .80 as useful in drawing "highly tentative and cautious" conclusions. In the example above, the alpha of .278 does not suggest acceptable agreement, and the researchers would therefore want to reevaluate operational definitions before attempting an additional reliability analysis. The following section examines interrater reliability testing for nominal data when more than two raters are involved.

Analysis of Nominal Data with Multiple Raters

Fleiss's Generalized Kappa

In many cases, more than two researchers participate in tests of interrater agreement with nominal data (Fleiss 1971, Light 1971). To facilitate tests with multiple observers, Fleiss (1971, 378) generalized the kappa procedure "to the measurement of agreement among any constant number of raters where there is no connection between the raters judging the various subjects." Thus, like Cohen (1960), Fleiss (1971) assumed independence among raters as well as mutually exclusive and exhaustive variable categories. Conceptually, the formula for Fleiss's generalized kappa coefficient is largely the same as the one used in calculating Scott's pi, Cohen's kappa, and Krippendorff's alpha. In generalized kappa calculations, which follow, N refers to the total number of subjects; n refers to the number of ratings per subject; and k indicates the number of categories to which assignments may be made. Subscript i denotes subjects ($i = 1, \ldots, N$); subscript j indicates categories ($j = 1, \ldots, k$); and n_{ij} refers to the number of raters assigned to ith subject in the jth category (Fleiss 1971).

To demonstrate generalized kappa, one might consider a hypothetical valence study in which $n = 5$ raters have evaluated $N = 20$ images, classifying each into one of $k = 3$ categories (positive, negative, neutral). Each row in Table 11.2, which displays the data for this analysis, indicates where each of the five raters classified an image. As an example, three raters considered the first image positive, no raters considered it negative, and two saw it as neutral. Following Fleiss (1971), the first task is to calculate values for p_1, p_2, and p_3, which will then indicate the proportion of all assignments made to each variable category (i.e., positive, negative, and neutral). The formula follows:

$$p_j = \frac{1}{Nn} \sum_{i=1}^{N} n_{ij}$$

Recalling the number of raters (5) and the number of subjects (20), a total of 100 assignments (ratings) appear in Table 11.2, with 43 positive assignments, 31 negative, and 26 neutral. As indicated in the table, the (column) totals result in the following proportions: $p_1 = .43$, $p_2 = .31$, and $p_3 = .26$. As an example of proportion calculations, p_1 resulted from the following equation:

$$p_1 = \frac{1}{100}(43) = .43$$

In generalized kappa tests, calculations of row agreement proportions follow calculations for column agreement proportions. This series of calculations,

Table 11.2   Hypothetical data for Fleiss kappa calculation

		Categories		
Image	*Positive (j =1)*	*Negative (j =2)*	*Neutral (j =3)*	P_i
1	3	0	2	.40
2	4	1	0	.60
3	1	1	3	.30
4	3	1	1	.30
5	0	4	1	.60
6	5	0	0	1.0
7	2	1	2	.20
8	0	3	2	.40
9	1	0	4	.60
10	3	2	0	.40
11	0	5	0	1.0
12	2	2	1	.20
13	4	0	1	.60
14	1	1	3	.30
15	3	2	0	.40
16	2	0	3	.40
17	0	4	1	.60
18	4	1	0	.60
19	5	0	0	1.0
20	0	3	2	.40
				$\Sigma P_i = \overline{10.3}$
Total	43	31	26	100
p_j	.43	.31	.26	1.0

represented by the formula below (Fleiss 1971, 379), results in a proportion of agreeing pairs given $n\,(n-1)$ possible pairs for each image:

$$P_i = \frac{1}{n(n-1)} \sum_{i=1}^{k} n_{ij}\left(n_{ij}-1\right) = \frac{1}{n(n-1)}\left(\sum_{j=1}^{k} n_{ij^2} - n\right)$$

The calculation below applies to the first row of assigned ratings:

$$\frac{1}{5(5-1)}\left(3^2-3\right)+\left(0^2-0\right)+\left(2^2-2\right) = \frac{1}{20}(6)+(2) = .05(8) = .40$$

Agreement proportions, shown in the right-hand column of Table 11.2, reflect calculations made for each row. Proportions are then summed toward the calculation of a mean value. The formula for calculating a mean value follows:

$$\frac{1}{N} \sum_{i=1}^{N} P_i$$

In Table 11.2, agreement proportions totaled 10.3, and therefore the mean would be calculated as:

$$\frac{1}{20}(10.3) = .515$$

As with the original kappa statistic, an expected mean must also be calculated, as follows:

$$\sum_{j=1}^{k} p_{j^2}$$

For the data in Table 11.2, the expected mean would be calculated by adding $.43^2 + .31^2 + .26^2 = .3486$. The calculation for generalized kappa is thus:

$$\frac{.515 - .3486}{1 - .3486} = \frac{.1664}{.6514} = .255$$

Calculations for generalized kappa are tedious but not especially difficult. Unfortunately, while SPSS includes a measure for two-rater kappa, it does not contain the Fleiss (1971) test for multiple observers. Fleiss kappa calculators are available on the Internet, but they should be used with caution, perhaps for the purpose of verifying "traditional" calculations completed with pencil and paper. The chapter now addresses the analysis of ordinal data with two raters.

Analysis of Ordinal Data with Two Raters

Cohen's Weighted Kappa

In 1968, Cohen extended his earlier conception of kappa (Cohen 1960) to a weighted approach through which ordinal agreement measures could be calculated. Weighted kappa became an important measure in the social sciences, allowing researchers to move beyond unordered nominal categories to measures containing ordered observations. Given the capacity to weight observations, researchers could move from agreement-based questions such as *"Do* children behave aggressively following exposure to a violent cartoon?" to *"How* aggressively do children behave following exposure to a violent cartoon?" Cohen (1968, 216) used the following formula in discussing weighted kappa:

$$1 - \frac{\Sigma v_{ij} f_{oij}}{\Sigma v_{ij} f_{cij}}$$

where v_{ij} represents an assigned cell weight, f_{oij} represents an observed cell frequency, and f_{cij} represents chance-expected frequency. To demonstrate weighted kappa, Table 11.3 contains data extending the three-level unordered valence measure to a five-level ordinal measure. As with nominal data, ordered categories for two raters can be situated in a square table, with agreement displayed along the diagonal.[6]

In Table 11.3, raters agreed on 39 (36.4%) of 107 cases, disagreeing on 68 (63.6%). But to what extent did raters disagree, given ordered categories? Researchers can discover important clues by assigning weights to values based on the degree to which those values appear off the diagonal of complete agreement.

Writing in biostatistics, Norman and Streiner (2000) encouraged scholars to weight observations quadratically, as shown in Table 11.4. With quadratic weighting, values along the diagonal (i.e., values indicating complete agreement) receive no weight. Values located one level off the diagonal receive a weight of 1^2, while values located two levels off the diagonal receive a weight of 2^2. Values located three levels from agreement receive a weight of 3^2, and values removed by four levels receive a weight of 4^2. One thus observes weights of 0, 1, 4, 9, and 16 in the five-category agreement table.

To accommodate the focus on levels of disagreement, Norman and Streiner (2000, 220–221) observed the traditional formula for kappa:

$$\frac{P_o - P_e}{1 - P_e}$$

Table 11.3 Hypothetical data for weighted kappa calculation

		Observer 1					
		Very Negative	*Somewhat Negative*	*Neutral*	*Somewhat Positive*	*Very Positive*	*Marginal Totals*
Observer 2	Very Negative	8	5 (4.44)	3 (3.91)	2 (4.08)	1 (3.37)	19
	Somewhat Negative	4 (3.36)	8	3 (4.11)	3 (4.30)	2 (3.55)	20
	Neutral	3 (4.04)	6 (5.61)	7	6 (5.16)	2 (4.26)	24
	Somewhat Positive	2 (3.87)	4 (5.37)	5 (4.73)	7	5 (4.08)	23
	Very Positive	1 (3.53)	2 (4.91)	4 (4.32)	5 (4.51)	9	21
	Marginal Totals	18	25	22	23	19	107

Table 11.4 Example of quadratic weights applied to cells

	Very Negative	Somewhat Negative	Neutral	Somewhat Positive	Very Positive
Very Negative	0	1	4	9	16
Somewhat Negative	1	0	1	4	9
Neutral	4	1	0	1	4
Somewhat Positive	9	4	1	0	1
Very Positive	16	9	4	1	0

The statisticians then redefined kappa by substituting $Q = (1 - P)$ for everything in the formula, thereby revising the equation as follows:

$$\frac{Q_e - Q_o}{Q_e} = 1 - \frac{Q_o}{Q_e}$$

Given that formula, one must calculate and sum weighted observed and expected disagreements. As with chi-square, expected values can be calculated by multiplying related marginals and dividing by the table total. Formulas for the summation of weighted observed and expected disagreements follow:

$$Q_o = \Sigma w_{ij} \times P_{oij}$$

$$Q_e = \Sigma w_{ij} \times P_{eij}$$

where $i\,j$ refers to off-diagonal cells and w_{ij} are the respective cell weights. Returning to Table 11.3, observed and expected weighted proportions can be calculated in the following manner:

$$
\begin{aligned}
Q_o = 1 \times 5/107 + 4 \times 3/107 + 9 \times 2/107 + 16 \times 1/107 &= .477 \\
+ 1 \times 4/107 + 1 \times 3/107 + 4 \times 3/107 + 9 \times 2/107 &= .345 \\
+ 4 \times 3/107 + 1 \times 6/107 + 1 \times 6/107 + 4 \times 2/107 &= .299 \\
+ 9 \times 2/107 + 4 \times 4/107 + 1 \times 5/107 + 1 \times 5/107 &= .412 \\
+ 16 \times 1/107 + 9 \times 2/107 + 4 \times 4/107 + 1 \times 5/107 &= .515 \\
Q_o \quad &= 2.048
\end{aligned}
$$

$$
\begin{aligned}
Q_e = 1 \times 4.44/107 + 4 \times 3.91/107 + 9 \times 4.08/107 + 16 \times 3.37/107 &= 1.034 \\
1 \times 3.36/107 + 1 \times 4.11/107 + 4 \times 4.30/107 + 9 \times 3.55/107 &= 0.529 \\
4 \times 4.04/107 + 1 \times 5.61/107 + 1 \times 5.16/107 + 4 \times 4.26/107 &= 0.410 \\
9 \times 3.87/107 + 4 \times 5.37/107 + 1 \times 4.73/107 + 1 \times 4.08/107 &= 0.609 \\
16 \times 3.53/107 + 9 \times 4.91/107 + 4 \times 4.32/107 + 1 \times 4.51/107 &= 1.144 \\
Q_e \quad &= 3.726
\end{aligned}
$$

Having summed observed and expected disagreements over all cells, one can then calculate a value for weighted kappa: $1 - 2.048/3.726 = .450$. As Norman and Streiner (2000) pointed out, weighted kappa usually shows a higher value than the more rigid kappa statistic, as the former recognizes "near agreement" and the latter does not. In this case, unweighted kappa equaled .163, well below the weighted coefficient.

In communication, weighted kappa can prove useful when raters seek to assess agreement on subjective variables with ordered categories. Weighted kappa offers "partial credit" for ratings based on levels of removal from the diagonal. The following section considers the intraclass correlation coefficient, which statisticians have linked with weighted kappa.

Intraclass Correlation Coefficient

Readers may be familiar with the Pearson Product Moment Coefficient of Correlation, which analyzes association between two continuous measures. Pearson's R examines pairs of observations and indicates whether scores on one measure vary significantly with scores on another. A different type of measure, the intraclass correlation coefficient (ICC), developed by Fisher (1946), describes relationships among units within groups. While each variable in a Pearson analysis contains its own mean and standard deviation, the ICC statistic is based on pooled data (Bartko 1966, Muller and Buttner 1994).

Fisher (1946) explained the difference between *intraclass* and *interclass* correlations through a discussion of n' pairs of brothers. One approach to assessing correlations between brothers would be to divide them into two groups based on age. "If we proceed in this manner," Fisher (1946, 211–212) wrote, "we shall find the mean of the measurements of the elder brother, and separately that of the younger brother…The correlation so obtained, being that between two classes of measurements, is termed for distinctiveness an *interclass* correlation." In such a case, the researcher is privy to the age of each participant and can form separate groups accordingly.

But in cases where the researcher lacks information about the age of participants, calculating a common mean (and standard deviation) becomes necessary. The researcher thus forms a single class consisting of observational pairs (i.e., x_1, x'_1, x_2, x'_2 …, x_n, x'_n). As Fisher (1946, 212) explained, "The intraclass correlation…may be expected to give a more accurate estimate of the true value than does any of the possible interclass correlations derived from the same material, for we have used estimates of the mean and standard deviation founded on $2n'$ instead of n' values."

For purposes here, an ICC statistic can be used to describe how well the judgments of a sample of raters associate with the judgments of a larger population of observers. As Nunnally and Bernstein (1994, 279) noted, generalizability theory suggests that different measures of the same individual stand to vary based on actual differences as well as random error: "Two judges may disagree with each other because their judgments contain random measurement error... [but] they may also differ because they respond to different attributes." Psychometrically, the intraclass correlation coefficient is expressed as:

$$\rho^2 = \frac{\sigma_{ind}^2}{\sigma_{ind}^2 + \sigma_{error}^2}$$

Where σ_{ind}^2 represents true variance and σ_{error}^2 constitutes error variance (see, for discussion, Nunnally and Bernstein 1994, 280).

Fleiss and Cohen (1973) showed that a weighted kappa value can be interpreted as an intraclass coefficient for a two-way analysis of variance, given the assumption that subjects and raters have been randomly selected (see, for discussion, Banerjee et al. 1999, 7). In fact, as Norman and Streiner (2000, 221) pointed out, when kappa has been weighted quadratically, as it has been in the present chapter, the two measures show exact equivalence (see also, Fletcher, Mazzi, and Nuebling 2011).[7] Moreover, the ICC statistic is readily available in SPSS and other statistical software packages.

A key concern in using an intraclass correlation coefficient as a measure of agreement is the selection of the correct ICC statistic. Six different forms of the ICC exist (Shrout and Fleiss 1979, McGraw and Wong 1996), and researchers must carefully select among the options. As the current chapter will demonstrate in its section on SPSS, the ICC statistic based on a two-way analysis of variance with random effects equals the value of a weighted kappa coefficient.

Analysis of Ordinal Data with Multiple Raters

Continuing from the previous section, the intraclass correlation coefficient serves as a viable option for testing agreement when more than two raters assess ordinal content. As Norman and Streiner (2000, 222) explained, the ICC can be reported as an average kappa score, eliminating the need to calculate weighted kappa coefficients two raters at a time. In addition to the ICC statistic, Kendall's coefficient of concordance, a nonparametric test, examines agreement among more than two raters.

Kendall's Coefficient of Concordance

When three or more individuals rank a set of items, Kendall's coefficient of concordance, also referred to as Kendall's *W* (Kendall and Smith 1939), can determine the extent to which raters agree on observations (Nussbaum 2015, 198). As Norman and Streiner (2000, 233–234) noted, agreement relates to the dispersion of individual mean ranks relative to the average mean rank: "This is analogous to the intraclass correlation coefficient, where agreement [is] captured in the variance between subjects."

Kendall's W coefficients range between 0 and 1, with 0 indicating no agreement and 1 indicating perfect agreement. To demonstrate the technique, Table 11.5 contains hypothetical ranks from three individuals, asked to consider the relative importance of 10 issues facing the nation. Following a similar table structure from Norman and Streiner (2000, 233), Table 11.5 also contains the average rank for each issue and the total sum of squared ranks for each issue as well. The following formula is one of two equations that can be used to calculate Kendall's coefficient of concordance:

$$W = \frac{12\Sigma R_j^2 - 3k^2 N (N+1)^2}{k^2 N (N^2 - 1)}$$

where R_j refers to the summed rank, N the number of subjects, and k the number of judges. The following equation contains data from Table 11.5; as indicated, it produces a W value of .230.

$$W = \frac{12(2893) - 3(3^2)10(11)^2}{3^2(10)(10^2 - 1)} = \frac{34716 - 32670}{8910} = \frac{2046}{8910} = .230$$

Nonparametric statistics often focus on statistical significance, and Norman and Streiner (2000, 234) provided the following formula for calculating a chi-square statistic associated with W:

$$\chi^2 = k(N-1)W$$

Table 11.5   Data for calculation of Kendall's W

Issue	*Voter 1*	*Voter 2*	*Voter 3*	*Mean Rank*	*Squared Sum*
Abortion	10	8	2	6.67	400
Deficit	8	6	4	6.00	324
Education	3	4	10	5.67	289
Gun Control	9	7	3	6.33	361
Health Care	1	3	9	4.33	169
Immigration	4	5	1	3.33	100
Income Tax	2	9	5	5.33	256
Trade	7	10	8	8.33	625
Transportation	6	2	7	5.00	225
Unemployment	5	1	6	4.00	144
					2,893

Applied to the current example, the formula calls for the following calculation: $3 \times 9 \times .230$. This calculation results in a chi-square-distributed value of 6.21 with $N - 1 = 9$ degrees of freedom. One would conclude here that the rankings are not significant and, accordingly, not concordant.

SPSS will calculate Kendall's W (Analyze > Nonparametric > k related measures), and web-based calculators are also available.[8] As with other Internet-based calculators, researchers should verify "reliability" with traditional calculations. The following section offers SPSS instructions for computing kappa and intraclass correlation coefficients.

Kappa Coefficient in SPSS

Like many statistical tests covered in this text, Cohen's kappa coefficient is available within the Crosstabs procedure in SPSS. As shown in Figure 11.1, data should be entered in column form, with units rated on the left margin and rater scores in columns to the right. After assigning row and column variables (i.e., raters), one should open the Statistics window and select Kappa. Once that selection is made, one should return to the Crosstabs window and click OK. Doing so will produce the statistical output shown in Table 11.6.

The Cases section in Table 11.6 reports the number of records analyzed ($N = 83$). That section is followed by a cross-tabulation similar to the one shown

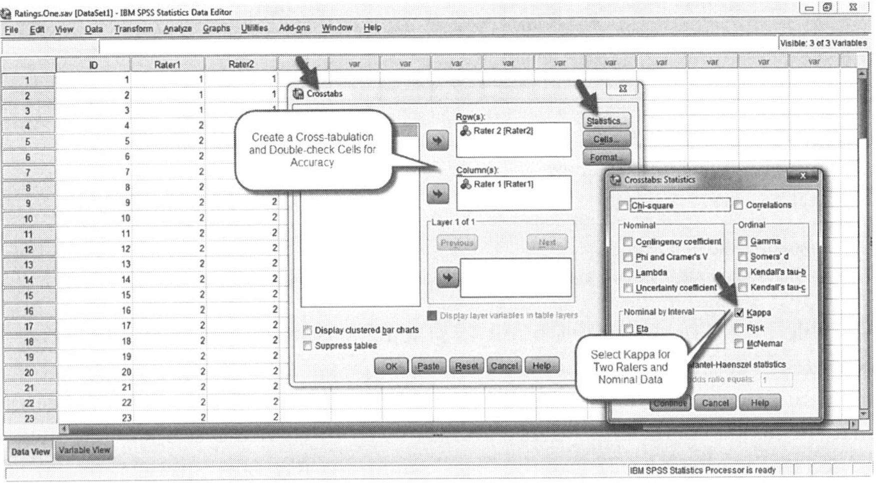

Figure 11.1 Screenshots for SPSS kappa analysis. Source: SPSS® Reprints Courtesy of International Business Machines Corporation, © 2014 International Business Machines Corporation

Table 11.6   SPSS output for kappa coefficient

		Cases				
	Valid		*Missing*		*Total*	
Rater 2 * Rater 1	N	Percent	N	Percent	N	Percent
	83	100.0%	0	.0%	83	100.0%

Rater 2 * Rater 1 Crosstabulation

			Rater 1			
			1	*2*	*3*	*Total*
Rater 2	1	Count	3	6	4	13
		% within Rater 2	23.1%	46.2%	30.8%	100.0%
		% within Rater 1	30.0%	9.5%	40.0%	15.7%
	2	Count	4	51	3	58
		% within Rater 2	6.9%	87.9%	5.2%	100.0%
		% within Rater 1	40.0%	81.0%	30.0%	69.9%
	3	Count	3	6	3	12
		% within Rater 2	25.0%	50.0%	25.0%	100.0%
		% within Rater 1	30.0%	9.5%	30.0%	14.5%
Total		Count	10	63	10	83
		% within Rater 2	12.0%	75.9%	12.0%	100.0%
		% within Rater 1	100.0%	100.0%	100.0%	100.0%

Symmetric Measures

		Value	Asymp. Std. Error[a]	Approx. T[b]	Approx. Sig.
Measure of Agreement	Kappa	.277	.088	3.379	.001
N of Valid Cases		83			

[a] Not assuming the null hypothesis.
[b] Using the asymptotic standard error assuming the null hypothesis.

in Table 11.1. For descriptive purposes, the table contains row and column percentages in addition to cell counts from the previous table. Following the crosstabulation, a value for kappa appears ($\kappa = .277$), and its p-value, $p < .001$, indicates that the kappa coefficient differs significantly from chance agreement. As mentioned earlier, though, .27 is in the "fair" range, and its relative conservatism may be due, in part, to the number of observations in the cell containing agreement on negative images. SPSS does not calculate Scott's pi or Krippendorff's alpha, but macros for the latter are available via the Internet (see Hayes and Krippendorff 2007). While kappa is widely used to measure rater agreement, it can be influenced by marginal distributions, and a second test can offer useful information.

Intraclass Correlation Coefficients in SPSS

Figure 11.2 contains screenshots for computing intraclass correlation coefficients in SPSS. The procedure can be accessed by selecting Analyze > Scale > Reliability Analysis. As with kappa, one should first select the raters to be assessed and then click the Statistics tab. This step, which opens the window shown to the right in Figure 11.2, allows the researcher to request Intraclass correlation coefficient. Importantly, it also allows the researcher to select Two-Way Random as the Model and Absolute Agreement as the Type; one should make certain these options are chosen before clicking Continue and then OK.

Table 11.7 contains output associated with the windows in Figure 11.2. The first section indicates that 107 cases were included in the analysis, consistent with the data shown earlier in Table 11.3. Next, although Cronbach's Alpha is included in the reliability analysis, it is not pertinent in this particular case. Statistics located under the heading Intraclass Correlation Coefficient are of principal interest, and as one observes in the row for Single Measures, the ICC statistic of .453 is consistent with the value calculated in the section addressing weighted kappa. Also included in Table 11.7 is a 95% confidence interval as well as F tests and significance levels. In this case, the ICC statistic is significant, indicating (moderate) agreement between the two coders. Should more than two raters be tested for agreement, an ICC statistic is still applicable as an indicator of ordinal agreement.

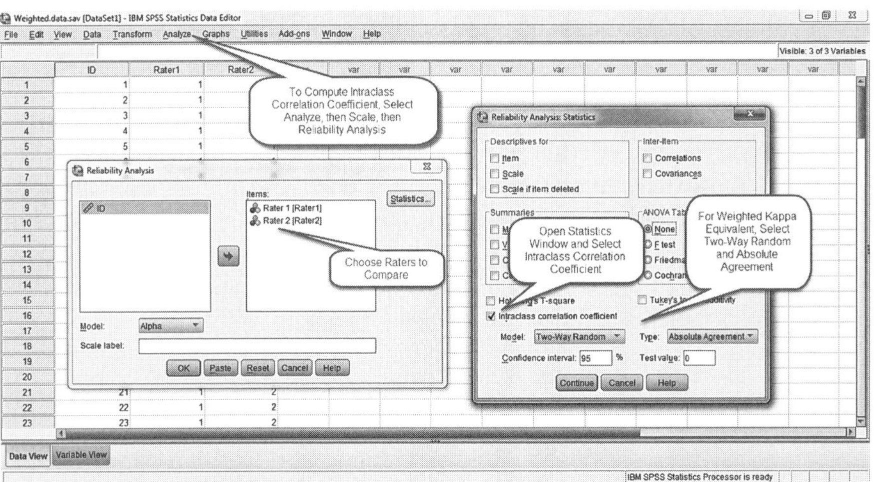

Figure 11.2 Screenshots for SPSS intraclass correlation coefficient. Source: SPSS® Reprints Courtesy of International Business Machines Corporation, © 2014 International Business Machines Corporation

Table 11.7   SPSS output for intraclass correlation coefficient

Case Processing Summary

		N	%
Cases	Valid	107	100.0
	Excluded[a]	0	.0
	Total	107	100.0

[a] Listwise deletion based on all variables in the procedure.

Reliability Statistics

Cronbach's Alpha	*N of Items*
.622	2

Intraclass Correlation Coefficient

	Intraclass Correlation[a]	*95% Confidence Interval*		*F Test with True Value 0*			
		Lower Bound	*Upper Bound*	*Value*	*df1*	*df2*	*Sig.*
Single Measures	.453[b]	.289	.592	2.647	106	106	.000
Average Measures	.624	.448	.744	2.647	106	106	.000

Two-way random effects model where both people effects and measure effects are random.
[a] Type A intraclass correlation coefficients using an absolute agreement definition.
[b] The estimator is the same, whether the interaction effect is present or not.

Chapter Summary

This chapter has focused on techniques for testing interrater agreement. At the nominal level, Scott's pi, Cohen's kappa, and Krippendorff's alpha test agreement between two raters, while Fleiss's kappa tests agreement when more than two raters assess data. At the ordinal level, Cohen's weighted kappa informs researchers of the extent to which two raters disagree. Intraclass correlation coefficients also provide indications of reliability with ordinal data, as does Kendal's coefficient of concordance. SPSS computes kappa and intraclass correlation coefficients and will also calculate Kendall's *W*.

Chapter Exercises

1. Using the following hypothetical data, which involves valence in photographic images, calculate Scott's pi and Cohen's kappa statistics. Then, use SPSS to calculate a kappa value, as shown in the chapter. Based on the Landis and Koch (1977a) benchmarks, how reliable were the observations of the two raters?

		Observer 1			
		Positive	*Negative*	*Neutral*	*Marginal*
Observer 2	Positive	7	4	6	17
	Negative	5	**40**	5	50
	Neutral	4	5	6	15
	Marginal	16	49	17	82

2. Using the following hypothetical data, which extends the valence of photographic images from three nominal levels to five ordinal categories, calculate a weighted kappa value, as shown in the chapter. Then, use SPSS to calculate an intraclass correlation coefficient. Are the two measures similar?

		Observer 1					
		Very Negative	*Somewhat Negative*	*Neutral*	*Somewhat Positive*	*Very Positive*	*Marginal Totals*
Observer 2	Very Negative	**8**	3	2	2	1	16
	Somewhat Negative	3	**9**	3	2	2	19
	Neutral	2	4	**8**	3	2	19
	Somewhat Positive	3	4	3	7	2	19
	Very Positive	2	2	3	3	7	17
	Marginal Totals	18	22	19	17	14	90

Notes

1 Generally, communication researchers code between 10% and 15% of a given sample to determine whether a satisfactory level of agreement has been reached.

2 Additionally, as Zwick (1988) noted, Scott's pi assumes homogeneity in rater marginals, and lacking such, a second interrater test (kappa or alpha) should be considered.

3 Cohen (1968) and other statisticians eventually developed weighted versions of κ to assess agreement for ordinal measures in addition to nominal variables (see, for discussion, Nelson and Pepe 2000).

4 Hayes and Krippendorff (2007) discussed alpha macros designed for SPSS and SAS. Readers should refer to their article for URLs where macros can be obtained.

5 This algebraic transformation applies to situations involving two raters.

6 As Cohen (1968, 213) explained, "In an agreement matrix, high reliability dictates that the values observed in k cells of the leading or agreement diagonal be higher

than the chance expectation dictated by the marginal values, and that, conversely, the off-diagonal cells representing disagreement have observed values which are smaller than those expected by chance."

7   In the context of dichotomous nominal data, Bloch and Kraemer (1989) demonstrated that kappa can be understood as an ICC statistic in a one-way analysis of variance equation. In proposing a different approach to kappa, the authors assumed raters followed the same underlying success rate (see also Landis and Koch 1977b).

8   One calculator appears at the following URL: https://www.statstodo.com/KendallW_Exp.php.

References

Agresti, Alan. 2007. *An Introduction to Categorical Data Analysis*, 2nd ed. Hoboken, NJ: John Wiley & Sons.

Banerjee, Mousumi, Michelle Capozzoli, Laura McSweeney, and Debajyoti Sinha. 1999. "Beyond Kappa: A Review of Interrater Agreement Measures." *Canadian Journal of Statistics*, 27: 3–23. DOI:10.2307/3315487.

Bartko, John J. 1966. "The Intraclass Correlation Coefficient as a Measure of Reliability." *Psychological Reports*, 19: 3–11. DOI:10.2466/pr0.1966.19.1.3.

Berelson, Bernard. 1952. *Content Analysis in Communication Research*. New York: Free Press.

Bloch, D. A., and H. C. Kraemer. 1989. "2 x 2 Kappa Coefficients: Measures of Agreement or Association." *Biometrics*, 45: 269–287.

Brennan, Robert L., and Dale J. Prediger. 1981. "Coefficient Kappa: Some Uses, Misuses, and Alternatives." *Educational and Psychological Measurement*, 41: 687–699. DOI:10.1177/001316448104100307.

Cohen, Jacob. 1960. "A Coefficient of Agreement for Nominal Scales." *Educational and Psychological Measurement*, 20: 37–46.

Cohen, Jacob. 1968. "Weighted Kappa: Nominal Scale Agreement with Provision for Scaled Disagreement or Partial Credit." *Psychological Bulletin*, 70: 213–220. DOI:10.1037/h0026256.

Craig, Robert T. 1981. "Generalization of Scott's Index of Intercoder Agreement." *Public Opinion Quarterly*, 45: 260–264. DOI:10.1086/268657.

Fisher, R. A. 1946. *Statistical Methods for Research Workers*, 10th ed. New York: Hafner.

Fleiss, Joseph L. 1971. "Measuring Nominal Scale Agreement among Many Raters." *Psychological Bulletin*, 76: 378–382. DOI:10.1037/h0031619.

Fleiss, Joseph L., and Jacob Cohen. 1973. "The Equivalence of Weighted Kappa and the Intraclass Correlation Coefficient as Measures of Reliability." *Educational and Psychological Measurement*, 33: 613–619. DOI:10.1177/001316447303300309.

Fleiss, Joseph L., Bruce Levin, and Myunghee Cho Paik. 2003. *Statistical Methods for Rates and Proportions*, 3rd ed. New York: John Wiley & Sons.

Fletcher, Ian, Mariangela Mazzi, and Matthias Nuebling. 2011. "When Coders are Reliable: The Application of Three Measures to Assess Inter-Rater Reliability/ Agreement with Doctor–Patient Communication Data Coded with the VR-

CoDES." *Patient Education and Counseling*, 82: 341–345. DOI:10.1016/j.pec.2011.01.004.

Hayes, Andrew F. 2005. *Statistical Methods for Communication Science*. Mahwah, NJ: Erlbaum.

Hayes, Andrew F., and Klaus Krippendorff. 2007. "Answering the Call for a Standard Reliability Measure for Coding Data." *Communication Methods and Measures*, 1: 77–89. DOI:10.1080/19312450709336664.

Holsti, Ole R. 1969. *Content Analysis for the Social Sciences and Humanities*. Reading, MA: Addison-Wesley.

Kendall, M. G., and B. Babington Smith. 1939. "The Problem of *m* Rankings." *Annals of Mathematical Statistics*, 10: 275–287.

Krippendorff, Klaus. 1980. *Content Analysis: An Introduction to Its Methodology*. Beverly Hills, CA: Sage.

Krippendorff, Klaus. 2004. "Reliability in Content Analysis: Some Common Misconceptions and Recommendations." *Human Communication Research*, 30: 411–433. DOI:10.1111/j.1468-2958.2004.tb00738.x.

Krippendorff, Klaus. 2008. "Systematic and Random Disagreement and the Reliability of Nominal Data." *Communication Methods and Measures*, 2: 323–338. DOI: 10.1080/19312450802467134.

Krippendorff, Klaus. 2011. "Agreement and Information in the Reliability of Coding." *Communication Methods and Measures*, 5: 93–112. DOI:10.1080/19312458.2011.568376.

Landis, J. Richard, and Gary G. Koch. 1977a. "The Measurement of Observer Agreement for Categorical Data." *Biometrics*, 33: 159–174.

Landis, J. Richard, and Gary G. Koch. 1977b. "A One-Way Components of Variance Model for Categorical Data." *Biometrics*, 33: 671–679.

Lang, Annie, Satoko Kurita, Bridget R. Rubenking, and Robert F. Potter. 2011. "MiniMAM: Validating a Short Version of the Motivation Activation Measure." *Communication Methods and Measures*, 5: 146–162. DOI:10.1080/19312458.2011.568377.

Lawal, Bayo. 2003. *Categorical Data Analysis with SAS and SPSS Applications*. Mahwah, NJ: Erlbaum.

Light, Richard J. 1971. "Measures of Response Agreement for Qualitative Data: Some Generalizations and Alternatives." *Psychological Bulletin*, 76: 365–377. DOI:10.1037/h0031643.

Lombard, Matthew, Jennifer Snyder-Duch, and Cheryl Campanella Bracken. 2002. "Content Analysis in Mass Communication: Assessment and Reporting of Intercoder Reliability." *Human Communication Research*, 28: 587–604. DOI:10.1111/j.1468–2958.2002.tb00826.x.

Lombard, Matthew, Jennifer Snyder-Duch, and Cheryl Campanella Bracken. 2004. "A Call for Standardization in Content Analysis Reliability." *Human Communication Research*, 30: 434–437. DOI:10.1111/j.1468-2958.2004.tb00739.x.

McGraw, Kenneth O., and S. P. Wong. 1996. "Forming Inferences about Some Intraclass Correlation Coefficients." *Psychological Methods*, 1: 30–46. DOI:10.1037/1082–989X.1.1.30.

Mitchell, Sandra K. 1979. "Interobserver Agreement, Reliability, and Generalizability of Data Collected in Observational Studies." *Psychological Bulletin*, 86: 376–390. DOI: 10.1037/0033-2909.86.2.376.

Muller, Reinhold, and Petra Buttner. 1994. "A Critical Discussion of Intraclass Correlation Coefficients." *Statistics in Medicine*, 13: 2465–2476. DOI:10.1002/sim.4780132310.

Nelson, Jennifer C., and Margaret S. Pepe. 2000. "Statistical Description of Interrater Variability in Ordinal Ratings." *Statistical Methods in Medical Research*, 9: 475–496. DOI:10.1177/096228020000900505.

Neuendorf, Kimberly A. 2002. *The Content Analysis Guidebook*. Thousand Oaks, CA: Sage.

Norman, Geoffrey R., and David L. Streiner. 2000. *Biostatistics: The Bare Essentials*, 2nd ed. Hamilton, Ontario, Canada: B. C. Decker.

Nunnally, Jum C., and Ira H. Bernstein. 1994. *Psychometric Theory*. 3rd ed. New York: McGrawHill, Inc.

Nussbaum, E. Michael. 2015. *Categorical and Nonparametric Data Analysis*. New York: Routledge.

Popping, Roel. 2010. "Some Views on Agreement to be Used in Content Analysis Studies." *Quality and Quantity*, 44: 1067–1078. DOI:10.1007/s11135-009-9258-3.

Potter, W. James, and Deborah Levine-Donnerstein. 1999. "Rethinking Validity and Reliability in Content Analysis." *Journal of Applied Communication Research*, 27: 258–284. DOI:10.1080/00909889909365539.

Riffe, Daniel, Stephen Lacy, and Frederick G. Fico. 2005. *Analyzing Media Messages: Using Quantitative Content Analysis in Research*, 2nd ed. Mahwah, NJ: Erlbaum.

Rodriguez, Lulu, and Ruby Lynn Asoro. 2012. "Visual Representations of Genetic Engineering and Genetically Modified Organisms in the Online Media." *Visual Communication Quarterly*, 19: 232–245. DOI:10.1080/15551393.2012.735585.

Scott, William A. 1955. "Reliability of Content Analysis: The Case of Nominal Scale Coding." *Public Opinion Quarterly*, 19: 321–325. DOI:10.1086/266577.

Shrout, Patrick E., and Joseph L. Fleiss. 1979. "Intraclass Correlations: Uses in Assessing Rater Reliability." *Psychological Bulletin*, 86: 420–428. DOI:10.1037/0033-2909.86.2.420.

Stokes, Maura E., Charles S. Davis, and Gary G. Koch. 2012. *Categorical Data Analysis Using SAS*, 3rd ed. Cary, NC: SAS Institute.

Tang, Wan, Hua He, and Xin M. Tu. 2012. *Applied Categorical and Count Data Analysis*. Boca Raton, FL: CRC Press.

Zwick, Rebecca. 1988. "Another Look at Interrater Agreement." *Psychological Bulletin*, 103: 374–378. DOI:10.1037/0033-2909.103.3.374.

12

Concluding Communication

I consulted many scholarly journals while writing this text, and in the spirit of *Communications in Statistics*, I offer some concluding thoughts about the use of categorical statistics in communication research. As a brief review, this book began with an introduction to categorical statistics and then moved to chi-square analysis, covering both goodness of fit and independence testing, before addressing measures of association. Next came a discussion of contingency tables in three dimensions, followed by two chapters on log-linear modeling and three chapters on logistic regression analysis. Additional chapters addressed probit models, Poisson and negative binomial regression, and interrater reliability analysis. A series of sidebars focused on measurement issues in categorical statistics.

Communication researchers have not used the techniques covered in this text as frequently as researchers in other social-scientific disciplines have, even though communication research methods lend themselves directly to categorical data analysis. Unlike the scholarly literature in disciplines such as political science, psychology, and sociology, communication research includes hundreds of studies based on content analysis (Riffe and Freitag 1997, Lovejoy et al. 2014), and with the emergence of social media (e.g., Facebook, Instagram, Twitter), content studies will almost certainly remain a popular form of research. In assigning numbers to texts – that is, in conducting content analyses – researchers typically create multiple categorical content measures, often using chi-square analysis to compare observed and expected frequencies. Some of these studies might benefit from a more advanced approach. Log-linear models, for instance, can reveal associations that may go unnoticed in bivariate analyses, and logit log-linear analysis allows multiple categorical variables to predict cell frequencies in one or more categorical response measures. Logistic regression allows continuous measures to function as predictors, and Poisson and negative binomial regression facilitate analyses of count data. These techniques help to control Type I error and, in the case of log-linear models, the procedures assist researchers in locating the most parsimonious representation of variable relationships. Bivariate chi-square tests generally lack this capacity, and like nonparametric techniques, in general, they do not produce parameter estimates.

Modeling techniques for categorical data do produce parameter estimates, and because they are designed specifically for nominal and ordinal measures, the procedures do not risk major assumption violations. In analyses with ordinal response measures, for instance, ordinal logistic regression allows researchers to avoid problems with unequal intervals, which can occur when individuals assign different meanings to vague quantifiers. Borrowing from Cliff (1996, 331), researchers can respond appropriately to "ordinal questions with ordinal data using ordinal statistics." This is not to suggest that techniques such as ordinary least squares (OLS) regression and the analysis of variance (ANOVA) should *never* be used with ordinal dependent measures. As indicated in the text, OLS regression and ANOVA may function reasonably well when response variables contain at least four levels; however, when dependent measures contain fewer than four levels or do not meet assumptions of normality, ordinal and multinomial regression techniques may offer more reliable estimates – and more useful information.

Additionally, in analyses containing dichotomous dependent measures, binary logistic regression models (and binary probit models) allow researchers to avoid pitfalls associated with linear probability models (Aldrich and Nelson 1984), ensuring appropriate floor and ceiling restrictions, while Poisson and negative binomial regression techniques generally prove superior in analyses of count data; traditional regression analyses may predict negative values, which, of course, are fundamentally impossible for counts, as measured across time.

Like techniques designed for analyses of interval-level data, tests and models developed for nominal and ordinal measures are also expected to meet certain assumptions. As this text suggested in its chapter on ordinal logistic regression, researchers should pay close attention to assumption violations, as statistical problems can inform research questions. If, for example, an ordinal logistic regression model does not show parallel lines, thus removing justification for a common odds ratio, what does the assumption violation reveal about the data at different variable levels? Is there a substantive reason for differential effects? Researchers should also monitor multivariate data analyses for the presence of zero-count cells, using delta (in SPSS) when empty cells emerge. Problems with empty cells often appear when samples are too small for the number of variables included in a study.

As discussed in the chapter addressing interrater reliability measures, cell frequencies can also play a role in analyses of agreement. When a given content measure does not contain a sufficient number of observations in each of its categories, with a preponderance of observations instead assigned to a single classification, agreement between coders may not exceed chance. Measures such as Cohen's kappa, Scott's pi, and Krippendorff's alpha account for chance agreement, as do other measures addressed in the chapter on interrater reliability. Researchers who analyze content should select one of these measures based on the level of data and the number of available coders.

As a final point of discussion, the first chapter in this text pointed out that research methodologist Paul Lazarsfeld, whose studies on voter behavior played a significant role in the development of communication theory and quantitative research methods, did not test relationships for statistical significance. Recognizing the potential influence of "correlated biases" (Selvin 1957), Lazarsfeld and his coauthors explained findings in terms of frequencies, using tables and figures to illustrate patterns. Had multivariate techniques for the analysis of categorical data been available, Lazarsfeld might have incorporated the procedures in his research.

In the twenty-first century, "statistical significance" has become a virtual requirement for publication of quantitative research (Levine et al. 2008, Levine 2013), even though attaining it often requires little more than a large sample (Leahey 2005). This is especially true of chi-square analysis and is one reason why measures of association need to accompany indicators of statistical significance. Techniques such as log-linear modeling and logistic regression analysis use odds ratios as measures of association, and in analyses based on large samples, these measures offer information that significance tests do not. Odds ratios inform researchers of whether statistically significant findings offer a degree of practical significance, and associated confidence intervals inform scholars of whether perceived associations are actually independent (i.e., contain the value 1.0). These indicators add a dose of conservatism to statistical analyses, making research more precise and informative. As this book has suggested, categorical statistics, in general, can add precision to quantitative research, informing scholars about conceptual processes in communication.

References

Aldrich, John H., and Forrest D. Nelson. 1984. *Linear Probability, Logit, and Probit Models.* Newbury Park, CA: Sage.

Cliff, Norman. 1996. "Answering Ordinal Questions with Ordinal Data Using Ordinal Statistics." *Multivariate Behavioral Research*, 31: 331–350.

Leahey, Erin. 2005. "Alphas and Asterisks: The Development of Statistical Significance Testing in Sociology." *Social Forces*, 84: 1–24.

Levine, Timothy R. 2013. "A Defense of Publishing Nonsignificant (ns) Results." *Communication Research Reports*, 30: 270–274. DOI:10.1080/08824096.2013. 806261.

Levine, Timothy R., Rene Weber, Craig Hullett, Hee Sun Park, and Lisa L. Massi Lindsay. 2008. "A Critical Assessment of Null Hypothesis Significance Testing in Quantitative Communication Research." *Human Communication Research*, 34: 171–187. DOI: 10.1111/j.1468–2958.2008.00317.x.

Lovejoy, Jennette, Brendan Watson, Stephen Lacy, and Daniel Riffe. 2014. "Assessing the Reporting of Reliability in Published Content Analyses: 1985–2010." *Communication Methods and Measures*, 8: 207–221. DOI:10.1080/19312458.2 014.937528.

Riffe, Daniel, and Alan Freitag. 1997. "A Content Analysis of Content Analyses: Twenty-Five Years of *Journalism Quarterly*." *Journalism & Mass Communication Quarterly*, 74: 515–524.

Selvin, Hanan C. 1957. "A Critique of Tests of Significance in Survey Research." *American Sociological Review*, 23: 519–527.

Appendix A

Chi-Square Table

			Right-Tail Probability			
df	*.100*	*.050*	*.025*	*.010*	*.005*	*.001*
1	2.706	3.841	5.024	6.635	7.879	10.828
2	4.605	5.991	7.378	9.210	10.597	13.816
3	6.251	7.815	9.348	11.345	12.838	16.266
4	7.779	9.488	11.143	13.277	14.860	18.467
5	9.236	11.070	12.833	15.086	16.750	20.515
6	10.645	12.592	14.449	16.812	18.548	22.458
7	12.017	14.067	16.013	18.475	20.278	24.322
8	13.362	15.507	17.535	20.090	21.955	26.124
9	14.684	16.919	19.023	21.666	23.589	27.877
10	15.987	18.307	20.483	23.209	25.188	29.588
11	17.275	19.675	21.920	24.725	26.757	31.264
12	18.549	21.026	23.337	26.217	28.300	32.909
13	19.812	22.362	24.736	27.688	29.819	34.528
14	21.064	23.685	26.119	29.141	31.319	36.123
15	22.307	24.996	27.488	30.578	32.801	37.697
16	23.542	26.296	28.845	32.000	34.267	39.252
17	24.769	27.587	30.191	33.409	35.718	40.790
18	25.989	28.869	31.526	34.805	37.156	42.312
19	27.204	30.144	32.852	36.191	38.582	43.820
20	28.412	31.410	34.170	37.566	39.997	45.315
21	29.615	32.671	35.479	38.932	41.401	46.797
22	30.813	33.924	36.781	40.289	42.796	48.268

(*Continued*)

(Continued)

	Right-Tail Probability					
df	*.100*	*.050*	*.025*	*.010*	*.005*	*.001*
23	32.007	35.172	38.076	41.638	44.181	49.728
24	33.196	36.415	39.364	42.980	45.559	51.179
25	34.382	37.652	40.646	44.314	46.928	52.620
26	35.563	38.885	41.923	45.642	48.290	54.052
27	36.741	40.113	43.195	46.963	49.645	55.476
28	37.916	41.337	44.461	48.278	50.993	56.892
29	39.087	42.557	45.722	49.588	52.336	58.301
30	40.256	43.773	46.979	50.892	53.672	59.703
40	51.805	55.758	59.342	63.691	66.766	73.402
50	63.167	67.505	71.420	76.154	79.490	86.661
60	74.397	79.082	83.298	88.379	91.952	99.607
70	85.527	90.531	95.023	100.425	104.215	112.317
80	96.578	101.879	106.629	112.329	116.321	124.839
90	107.565	113.145	118.136	124.116	128.299	137.208
100	118.498	124.342	129.561	135.807	140.169	149.449

Appendix B

SPSS Code for Selected Procedures

Chapter 2: Goodness of Fit

```
*Nonparametric Tests: One Sample.
NPTESTS
  /ONESAMPLE TEST (Sex)
    CHISQUARE(EXPECTED = CUSTOM(CATEGORIES = 1 2
    FREQUENCIES = 0.421 0.579))
  /MISSING SCOPE = ANALYSIS USERMISSING = EXCLUDE
  /CRITERIA ALPHA = 0.05 CILEVEL = 95.
```

Chapter 3: Three-Way Contingency Table (Horseracing)

```
CROSSTABS
  /TABLES = Period BY Drug BY NP
  /FORMAT = AVALUE TABLES
  /STATISTICS = CHISQ RISK CMH(1)
  /CELLS = COUNT ROW
  /COUNT ROUND CELL.
```

Chapter 3: Three-Way Contingency Table

```
CROSSTABS
  /TABLES = V081101 BY V085108 BY V081102
  /FORMAT = AVALUE TABLES
  /STATISTICS = RISK CMH(1)
```

```
/CELLS = COUNT ROW
/COUNT ROUND CELL.
```

Chapter 4: Three-Factor Log-linear Model

```
GENLOG V081101 V081102 V085084A
  /MODEL = MULTINOMIAL
  /PRINT = FREQ RESID ADJRESID ZRESID DEV ESTIM CORR COV
  /PLOT = NONE
  /CRITERIA = CIN(95) ITERATE(20) CONVERGE(0.001) DELTA(.5)
  /DESIGN V081101 V081102 V085084A V081102*V085084A.
```

Chapter 4: Three-Factor Ordinal Log-linear Model (2 interactions)

```
GENLOG V6218 V6478 V6253 WITH A B
  /MODEL = MULTINOMIAL
  /PRINT = FREQ RESID ADJRESID ZRESID DEV ESTIM CORR COV
  /PLOT = NONE
  /CRITERIA = CIN(95) ITERATE(20) CONVERGE(0.001) DELTA(.5)
  /DESIGN V6218 V6478 V6253 A B.
```

Chapter 4: Four-Factor Log-linear Model with Covariate

```
GENLOG V081101 V081102 V085061 V085062 WITH V085109
  /MODEL = MULTINOMIAL
  /PRINT = FREQ RESID ADJRESID ZRESID DEV ESTIM CORR COV
  /PLOT = RESID(ADJRESID) NORMPROB(ADJRESID)
  /CRITERIA = CIN(95) ITERATE(20) CONVERGE(0.001) DELTA(.5)
  /DESIGN V081101 V081102 V085061 V085062 V085109
    V085061*V085062 V081101*V081102.
```

Chapter 5: Three-Factor Logit Log-linear Model with Covariate

```
GENLOG V085062 BY V081101 V081102 V085061 WITH V085109
  /MODEL = MULTINOMIAL
  /PRINT = FREQ RESID ADJRESID ZRESID DEV
  /PLOT = NONE
```

```
/CRITERIA=CIN(95) ITERATE(20) CONVERGE(0.001) DELTA(.5)
/DESIGN V085062 V085062*V081101 V085062*V081102 V085062
  *V085061 V085062*V085109.
```

Chapter 6: Binary Logistic Regression Model with Categorical Predictors

```
LOGISTIC REGRESSION VARIABLES V6142
  /METHOD=ENTER V6150 V6151
  /METHOD=ENTER V6575
  /METHOD=ENTER V6104
  /CONTRAST (V6104)=Simple(1)
  /CONTRAST (V6575)=Simple(1)
  /CONTRAST (V6150)=Indicator
  /CONTRAST (V6151)=Indicator(1)
  /PRINT=GOODFIT CORR CI(95)
  /CRITERIA=PIN(0.05) POUT(0.10) ITERATE(20) CUT(0.5).
```

Chapter 7: Multinomial Model with Categorical Predictors

```
NOMREG V6520 (BASE=LAST ORDER=ASCENDING) BY V6150
V6345 V6575
  /CRITERIA CIN(95) DELTA(0) MXITER(100) MXSTEP(5)
    CHKSEP(20) LCONVERGE(0) PCONVERGE(0.000001)
    SINGULAR(0.00000001)
  /MODEL
  /STEPWISE=PIN(.05) POUT(0.1) MINEFFECT(0) RULE(SINGLE)
    ENTRYMETHOD(LR) REMOVALMETHOD(LR)
  /INTERCEPT=INCLUDE
  /PRINT=CLASSTABLE FIT PARAMETER SUMMARY LRT CPS STEP
    MFI.
```

Chapter 8: Ordinal Regression Model with Categorical Predictors

```
PLUM V085109 BY V081101 V083097 V083024B
  /CRITERIA=CIN(95) DELTA(0) LCONVERGE(0) MXITER(100)
    MXSTEP(5) PCONVERGE(1.0E-6) SINGULAR(1.0E-8)
  /LINK=LOGIT
  /PRINT=CELLINFO FIT PARAMETER SUMMARY TPARALLEL.
```

Chapter 9: Binary Probit Analysis

```
* Generalized Linear Models.
  GENLIN V085108 (REFERENCE = LAST) BY V081101 V081102
    V085062 (ORDER = ASCENDING)
  /MODEL V081101 V081102 V085062 INTERCEPT = YES
  DISTRIBUTION = BINOMIAL LINK = PROBIT
  /CRITERIA METHOD = FISHER(1) SCALE = 1 COVB = MODEL
    MAXITERATIONS = 100 MAXSTEPHALVING = 5
    PCONVERGE = 1E-006(ABSOLUTE) SINGULAR = 1E-012
    ANALYSISTYPE = 3(WALD) CILEVEL = 95 CITYPE = WALD
    LIKELIHOOD = FULL
  /MISSING CLASSMISSING = EXCLUDE
  /PRINT CPS DESCRIPTIVES MODELINFO FIT SUMMARY
    SOLUTION.
```

Chapter 10: Poisson Regression

```
* Generalized Linear Models.
  GENLIN V083019 BY V081101 V081102 V085062
    (ORDER = ASCENDING)
  /MODEL V081101 V081102 V085062 INTERCEPT = YES
  DISTRIBUTION = POISSON LINK = LOG
  /CRITERIA METHOD = FISHER(1) SCALE = 1 COVB = MODEL
    MAXITERATIONS = 100 MAXSTEPHALVING = 5
    PCONVERGE = 1E-006(ABSOLUTE) SINGULAR = 1E-012
    ANALYSISTYPE = 3(WALD) CILEVEL = 95 CITYPE = WALD
    LIKELIHOOD = FULL
  /EMMEANS TABLES = V081101 SCALE = ORIGINAL
  /EMMEANS TABLES = V081102 SCALE = ORIGINAL
  /EMMEANS TABLES = V085062 SCALE = ORIGINAL
  /MISSING CLASSMISSING = EXCLUDE
  /PRINT CPS DESCRIPTIVES MODELINFO FIT SUMMARY
    SOLUTION
  /SAVE MEANPRED STDDEVIANCERESID.
```

Chapter 11: Kappa Test

```
CROSSTABS
  /TABLES = Rater2 BY Rater1
  /FORMAT = AVALUE TABLES
```

```
/STATISTICS = KAPPA
/CELLS = COUNT ROW COLUMN
/COUNT ROUND CELL.
```

Chapter 11: Intraclass Correlation Coefficient

```
GET
    FILE = 'C:\Users\admin\Desktop\Weighted.data.sav'.
    DATASET NAME DataSet1 WINDOW = FRONT.
    RELIABILITY
    /VARIABLES = Rater1 Rater2
    /SCALE('ALL VARIABLES') ALL
    /MODEL = ALPHA
    /ICC = MODEL(RANDOM) TYPE(ABSOLUTE) CIN = 95 TESTVAL = 0.
```

Index